Simplifying Service Management with Consul

Overcome connectivity and security challenges within dynamic service architectures

Robert E. Jackson

BIRMINGHAM—MUMBAI

Simplifying Service Management with Consul

Copyright © 2021 Packt Publishing

Group Product Manager: Wilson Dsouza
Publishing Product Manager: Rahul Nair
Senior Editor: Arun Nadar
Content Development Editor: Sayali Pingale
Technical Editor: Arjun Varma
Copy Editor: Safis Editing
Project Coordinator: Shagun Saini
Proofreader: Safis Editing
Indexer: Subalakshmi Govindhan
Production Designer: Vijay Kamble

First published: October 2021

Production reference: 1131021

Published by Packt Publishing Ltd.
Livery Place
35 Livery Street
Birmingham
B3 2PB, UK.

978-1-80020-262-7

www.packt.com

Contributors

About the author

Robert E. Jackson earned his bachelor of science in electrical engineering from Purdue University in 1997, and since then has found himself in a multitude of pre- and post-sales positions with various successful start-ups, mostly in the network access technology space. He has played multiple roles over the course of his career, including sales engineer, solutions engineer, integration engineer, network engineer, and product manager. Throughout all of these engineering positions he never learned to drive a train, but he was able to experience the digital transformation from traditional data centers to cloud computing from multiple viewpoints. He is currently employed at HashiCorp as a senior solutions engineer, based in the northeast area of the United States.

Special acknowledgements to Ancil McBarnett for his continued support and guidance, and for taking a chance on an old cable guy. I would also like to thank Patrick Roche and Gina Cocivera-Wright for their encouragement – you guys are the best cheerleaders. Last but certainly not least I would like to thank Rachael Jackson for the weekend, nights, and anxiety-ridden freak-outs throughout the creation of this book.

About the reviewer

Alex Schenck is a senior solutions engineer at HashiCorp. He holds a BSc in information technology, focusing on network engineering, from the New England Institute of Technology and an MSc in management from Worcester Polytechnic Institute. He has worked as an IT admin, tech support engineer, sales engineer, and solutions architect for a variety of organizations, including EMC and Zerto. In his spare time, he enjoys flying his 1978 Piper Archer II, learning German as a second language, and traveling. He lives in Rhode Island with his wife Melissa and two cats, Maid and Butler.

I would like to thank Rob for giving me the opportunity to be a part of his journey in creating this book. Reviewing his content gave me an opportunity to grow as a technologist, and you learn a lot about a person through their writing style. I am happy to count Rob as a friend and colleague at HashiCorp and I am honored to have taken this journey with him.

Thank you to Shagun at Packt as well, for her enormous patience and keeping me on track.

Ed Featherston is a Distinguished Technologist with more than thirty-nine years of designing and delivering **Enterprise Architecture (EA)** solutions for the most complex technical environments. Ed's experience has included consulting engagements with numerous Fortune 500 companies across a variety of industries, including financial services, government, pharmaceutical, and retail. He has significant expertise in systems integration, including internet and intranet, client/server, and middleware technologies. He's successfully evaluated current-state architecture and business and system requirements and then formulated and supported the architecture and integration strategy. Ed's used best practices and sound processes to rebuild IT organizations by integrating and aligning business applications and defining roles and responsibilities to achieve optimal business performance and technology functionality.

In recent years, Ed has also expanded his profile to become a leading voice on technology industry trends including cloud computing and blockchain. He regularly engages in industry conversations on social media channels like Twitter and routinely shares learnings, insights, and best practices with his 5,500 followers.

Table of Contents

3

Keep It Safe, Stupid, and Secure Your Cluster!

4

Data Center (Not Trade) Federation

Section 2: Use Cases Deep Dive

5

Little Bo Peep Lost Her Service

6

Connect Four or More

7

Animate Me

8

Where Do We Go Now?

Other Books You May Enjoy

Index

Preface

Have you ever spent hours troubleshooting a network, only to realize the network addresses for your applications changed? Or perhaps you've had to deflect flying coffee mugs from irate project managers because it is taking weeks to update a firewall or load balancer. Maybe your engineering teams are struggling with end-to-end encryption between their services, reinventing the wheel with every project, or leaving security up to the fictitious perimeter surrounding the network. If these situations sound familiar, HashiCorp Consul just might be the answer for your woes!

Consul provides a platform-independent solution for service discovery, network automation, and secure service communications. This book walks through the foundation of the Consul architecture, explaining how the servers and clients operate. We'll build our own Consul system in Amazon Web Services, and secure the system against cowans and eavesdroppers. With an established cluster, we'll start deploying services into the system, and show how Consul discovers and monitors these services. Multiple methods of sharing service information will be discussed. However, the continuing focus is on how Consul utilizes this information to dynamically adjust network infrastructure and facilitate secure connections among the registered services. Finally, we'll do a quick walk-through of HCP Consul, a Software as a Service solution offered on HashiCorp Cloud Platform.

Who this book is for

Any solution architects or DevOps engineers interested in the challenges and solutions of network management in a dynamic hybrid-cloud environment. A background in basic networking is helpful as we discuss device communication and connectivity, as well as the layers of the network stack. Knowledge of Terraform will be beneficial, as we'll be using Terraform to create and work with our system, but it is not required.

What this book covers

Chapter 1, Consul Overview – Operation and Use Cases, starts the book with a short overview of Consul and a high-level understanding of the three primary Consul use cases.

Chapter 2, Architecture – How Does It Work?, shows us how the servers and agents operate and communicate, now that we have a basic understanding of "why" Consul.

Chapter 3, Keep It Safe, Stupid, and Secure Your Cluster!, looks at securing the entire system, now that we understand how Consul operates. We'll learn how to apply Zero Trust to all of our applications.

Chapter 4, Data Center (Not Trade) Federation, looks at how Consul facilitates the extension of the cluster across multiple environments for world domination, now that, within the global and everchanging world, our applications must extend beyond the geographical borders of data centers.

Chapter 5, Little Bo Peep Lost Her Service, looks at how Consul not only discovers services – the foundation of Consul's operation being service discovery, but within those services, it also monitors and shares service status information.

Chapter 6, Connect Four or More, looks at, with knowledge of our services and their location(s), using Consul to identify and secure the communication among our services.

Chapter 7, Animate Me, covers how to simplify our networking infrastructure management with Consul's service information, with an entire world outside of Consul that can make use of its dynamic service discovery.

Chapter 8, Where Do We Go Now?, provides some pathways to continue your Consul education and operation. We introduce the HashiCorp certification program for Consul Associate Certification for further knowledge. Additionally, we'll show you how to sign up for HashiCorp's hosted Consul solution enabling the adoption of Consul without the overhead of managing the cluster itself.

To get the most out of this book

Throughout this book, we'll be using Terraform to build and manage our Consul system, and although knowledge of Terraform would be beneficial, it isn't required. We'll walk through the Terraform code and what it performs in order to create and manage the Consul system. Much of our operations within the book utilize the bash command line. Familiarity with common bash operations and the use of emacs or vi for editing text documents will be useful.

Software/hardware covered in the book	OS requirements
Consul	Windows, Linux, MacOS
Terraform	Windows, Linux, MacOS
Text editor of choice	

Download the color images

We also provide a PDF file that has color images of the screenshots/diagrams used in this book. You can download it here: `https://static.packt-cdn.com/downloads/9781800202627_ColorImages.pdf`.

Download the example code files

You can download the example code files for this book from GitHub at `https://github.com/PacktPublishing/Simplifying-Service-Management-with-Consul`.

In case there's an update to the code, it will be updated on the existing GitHub repository. We also have other code bundles from our rich catalog of books and videos available at `https://github.com/PacktPublishing/`. Check them out!

Conventions used

There are a number of text conventions used throughout this book.

`Code in text`: Indicates code words in text, database table names, folder names, filenames, file extensions, pathnames, dummy URLs, user input, and Twitter handles. Here is an example: "Within the Terraform output, you'll see a value for `CONSUL_HTTP_ADDR`."

A block of code is set as follows:

```
Apply complete! Resources: 23 added, 0 changed, 0 destroyed.

Outputs:

CONSUL_HTTP_ADDR = "http://18.218.165.4:8500"
Consul_Server_IPs = [
  "3.139.233.155",
```

```
    "18.218.165.4",
    "18.219.179.22",
]
```

Any command-line input or output is written as follows:

```
$ export AWS_ACCESS_KEY_ID=AKIA26CFBA32D6JFVJJB
$ export AWS_SECRET_ACCESS_KEY=NYaSFhAzZqOq4bqSGJda/Y22Guc…
```

Bold: Indicates a new term, an important word, or words that you see onscreen. For example, words in menus or dialog boxes appear in the text like this. Here is an example: "On the left side of the page, you should see a **Users** section where you can utilize an existing user or add a new one."

> **Tips or important notes**
> Appear like this.

Get in touch

Feedback from our readers is always welcome.

General feedback: If you have questions about any aspect of this book, mention the book title in the subject of your message and email us at customercare@packtpub.com.

Errata: Although we have taken every care to ensure the accuracy of our content, mistakes do happen. If you have found a mistake in this book, we would be grateful if you would report this to us. Please visit www.packtpub.com/support/errata, selecting your book, clicking on the Errata Submission Form link, and entering the details.

Piracy: If you come across any illegal copies of our works in any form on the Internet, we would be grateful if you would provide us with the location address or website name. Please contact us at copyright@packt.com with a link to the material.

If you are interested in becoming an author: If there is a topic that you have expertise in and you are interested in either writing or contributing to a book, please visit authors. packtpub.com.

Share Your Thoughts

Once you've read *Simplifying Service Management with Consul*, we'd love to hear your thoughts! Scan the QR code below to go straight to the Amazon review page for this book and share your feedback.

https://packt.link/r/1800202628

Your review is important to us and the tech community and will help us make sure we're delivering excellent quality content.

Section 1: Consul Use Cases and Architecture

Welcome to the consoling world of Consul. If you've ever built a structure, you know that the foundation is the most important part. If you don't believe me, just look at the leaning tower of Pisa! This section covers the foundational aspects of the Consul system, briefly reviewing why Consul is such a critical component in today's dynamic world, as well as how the Consul system operates to support various use cases.

This part of the book comprises the following chapters:

- *Chapter 1, Consul Overview – Operation and Use Cases*
- *Chapter 2, Architecture – How Does It Work?*
- *Chapter 3, Keep It Safe, Stupid, and Secure Your Cluster!*
- *Chapter 4, Data Center (Not Trade) Federation*

1
Consul Overview – Operation and Use Cases

Welcome to the wonderful world of **HashiCorp Consul**, a versatile utility to manage, automate, and securely connect all of the services within your network. Thanks for joining me on this ride, and I hope we'll both learn a lot throughout the process. Within this chapter, we're going to be reviewing Consul at a high level just to understand its basics and applications; essentially, we'll be digging the foundation and pouring the concrete to create a stable base on which to form your Consul structure. And if you've ever looked at the leaning tower of Pisa, you know that a solid foundation is critical for any important structure! Within this chapter, we'll be learning about the following:

- The Consul system – what are the components and how do they live together in harmony?

- Consul service discovery – touching on the foundational use case on which all of the other Consul functionality builds

- Consul service mesh – the basics of Consul service mesh to securely enable service-to-service communication
- Network automation – a high-level view of the application of service discovery to automate your network infrastructure

We won't be getting into any code snippets quite yet, but rest assured they are coming. If you're familiar with the concepts and use cases of Consul, go ahead and skip to those chapters – I promise I won't be hurt (too much). However, this would be an excellent chapter to hand your manager when they ask, *what in the world are you doing with this newfangled invention?*

The Consul system – servers and clients

Sometimes, in order to understand exactly how something works, we need to understand the components involved. After all, within every form of communication, there is a transmitter and a receiver, and if we aren't familiar with either one it can be difficult to understand the message. Consul, at its core, is a communications engine. It provides a new structure to facilitate the communications of network services, not just network devices. If you've spent any time in the networking field, you'll have heard terms such as IP address, subnet mask, and gateway address. These are all critical components involved in the communications of your network devices. However, in the glorious world of the cloud, microservices, and dynamic adjustments to the network, these components will change as the network waxes and wanes. Throughout this painful process, we've learned that the addressing system we've grown to know and love was simply a means to an end. That end is the connection and efficient communication of our services. So how does Consul do that? We are about to find out!

Consul servers – the master of their domain

Within almost any operational system or structure, there is a brain. This might not be the Brain who, along with his sidekick Pinky, attempts to take over the world day after day. We're focusing here on the brain that makes the decisions. The brain that determines whether to have Indian food or Thai food for dinner. The brain that figures out what to wear on a particular day. The brain that decides which employee gets the biggest raise. Now, at this point, you might be saying *computers and applications don't have brains; they only do what we tell them to do.* You are absolutely correct, but that doesn't mean that the machines don't have the logic and intelligence, even if artificial, to make decisions. In some ways, they are superior, as machines and applications are able to focus on logic and less on ego and emotion. Right, Mr. Spock? *That is correct, you subservient and primitive mammal.*

When contemplating the brain of Consul, we must look at the **Consul server cluster**. Looking at synonyms for *server*, we find *servant*, and that is exactly what the Consul server does for the distributed Consul clients (which we'll discuss later). The servers also work together, with an elected server making primary decisions, and distributing those decisions to the other servers within the overall cluster. So, what kind of decisions do these servants of my network make?

For starters, one of the most useful decisions the Consul server makes is where to find services. Everything that Consul does is predicated on the need to not only discover the services on the network, and their location, but also to share that information with any entitled entity asking for that information. If you think of the old telephone switchboards, somebody (an application) would pick up a phone and instantly be connected with a central operator (the server). The caller (application) would request a particular number (service) to be connected to, such as *Transylvania 6-5000*. The operator (server) would then move those giant plugs around and connect the caller (application) with the destination party (service).

Figure 1.1 – Telephone switchboard circa 1945

But what if I have applications that I don't want to be able to find specific services? Well, I'm glad you asked! In some cases, you have people or machines wishing to contact other people to talk to them about extending the warranty on their car. Perhaps I don't want to receive those messages. Well, I would tell that central operator to not allow those callers to contact me, kind of like a prehistoric *Do Not Call Registry*. These decisions are also managed by the Consul servers in accordance with your directive.

OK, it sounds like Consul can make a lot of cool and intelligent decisions automatically, but where does it get the information required to make those decisions? Well, how is any decision made? An employee's raise might be based upon their work performance considering they were also writing a book. What clothing we decide to wear is based on occasion and environment. What to eat for dinner can be based on...well, too many variables to list here! Every decision we make is only as good as the data we have. The Consul client provides that data.

> **Important Note**
>
> Within the official Consul documentation, there is a single *agent* that can be configured in both client and server mode. Throughout this text, we will refer to the server agents simply as the *server*, and the client agent simply as the *client*.

The cluster constituents

As the brain is the decision-maker in any system, there must be entities that will feed that data and perform the work based on that decision. For example, when the manager makes their decisions, the data is likely fed by board members, senior management executives, and hopefully team members. That being said, we all know the team members are the ones that do all of the work!

The plethora of data fed into the decisions made by the Consul server structure consists of two categories:

- Consul configuration
- Network data

The configuration of course is determined by those that are operating the Consul system, hopefully well thought out, defined as code, and peer-reviewed. The data, however, is fed by Consul clients that are distributed throughout the network. These clients may co-reside with the application, or they may travel alongside the application watching its every move. Regardless of the location, the client is not only a client for the server, but it is also a client for the application, representing the application and reporting information about the application to the server. This enables Consul's deployment within the network without the need to modify existing applications or services. Furthermore, a single client can represent multiple services and applications.

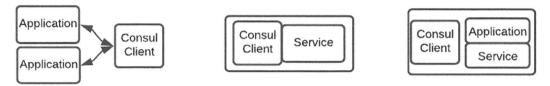

Figure 1.2 – Agent deployment options

> **Important Note**
> Care should be taken when utilizing an external service monitor with the Consul client. The data available to the client is reduced in this configuration and it is less flexible with respect to migrating services.

The Consul client not only monitors the applications that it is associated with, but it also collects information about the distance between itself and other clients within the network. This allows the servers to make more intelligent decisions about which instances of the applications or services are best suited to serve the impending request. The concept is very similar to emergency services when you dial for police or the fire service. There is one centralized number that you dial into (that is, 911 in the United States), but that centralized number connects you with whatever dispatch is closest to your current location. As Voice over IP services spread, this was a difficult challenge that increased network intelligence helped us overcome.

The marriage of server and client

OK, now that we know what the servers do, and we know what clients do, how does this entire system work together? As I've mentioned before, Consul is all about communication. The Consul servers need to communicate with each other, and the clients all need to communicate with the server cluster.

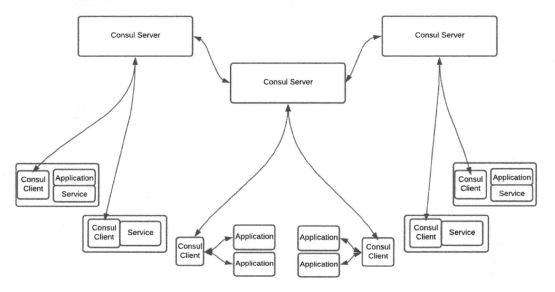

Figure 1.3 – Consul servers and agents living in harmony

The server communication is pretty manageable. It's a small dinner party with 3, 5, or 7 individuals. However, as you scale from tens to hundreds to potentially thousands of clients, having all of these clients report back to the relatively small server cluster can be quite overwhelming. Our intimate dinner party has expanded to a much larger party with tens, hundreds, or potentially thousands of guests. Can you imagine all of them wanting to talk to the head table at once! However, if the guests were able to chat amongst themselves, they could share their own information, discover new things, and only update the head table with pertinent and relevant information. In some cases, we might call this chat *gossiping*. Oddly enough, that's exactly the protocol Consul uses for this interaction, but we'll dig deeper into that in a later chapter.

Alright, now we have an understanding of the purpose of the servers, the purpose of the clients, and how they interact and communicate with each other. But what the heck can we do with this amazingly beautiful system? Why do I even care?

Oh where, oh where have my services gone?

As mentioned previously, one of the biggest problems related to the implementation of private and public cloud infrastructures is the fluidity of the network. Back in my day, any changes to the network required a myriad of steps. Let's focus on the simple need to deploy a new application. Let's assume that our application has already been written and tested:

1. Review the machine requirements requested.

2. Obtain the number of quotes for the requested machine(s).

3. Submit a purchase requisition.

4. Justify the purchase request with a number of management layers depending on cost.

5. Submit the purchase order to the vendor.

6. Wait for the machines to arrive.

7. Decide what IP address we're going to allocate.

8. Vote on what inventive name we'll call the machine(s).

9. Determine where in the warm and loud server room we're going to mount the machine(s).

10. Figure out how we're going to connect the machine(s).

11. Once the machine arrives, install the actual hardware and plug it in.

12. Argue with facilities about where to put the cardboard and leftover packing materials.

13. Configure the machine with the allocated network parameters (address and name).

14. Troubleshoot routing and address resolution tables.

15. Provide the application team with the address and credentials for the server.

Does any of this sound familiar? Each of these steps required human discussion (which is inherently unreliable) and usually paperwork and emails that got lost in the shuffle. Eventually, it got better with ticketing systems, but instead of getting buried in emails, we got buried in tickets that took days, weeks, and sometimes months to process. If our services needed to move to different servers or different areas of the network, we would have to start this entire process again. Now, as we've migrated toward more automation in the private and public clouds, let's look at how we've improved:

1. Review the machine requirements requested.
2. Identify what virtual network it needs to live in.
3. Execute the appropriate code modules to deploy the infrastructure.
4. Troubleshoot connectivity problems.
5. Provide the application team with the address and credentials for the server.

So many steps have disappeared, and the entire process of acquiring and deploying machines for our applications and services has been drastically simplified! However, the law of unintended consequences has not been eliminated. If you noticed, along with the elimination of cabling and mounting the hardware, we have also lost a lot of control surrounding the addressing of the devices. Of course, we can stipulate in what network our application or services should reside, but as we introduce new versions of the application, expand the application into other regions, or suffer a service loss and have to redeploy, the last thing we want to deal with is allocating and assigning addresses. But Rob, if I don't have an address, how do I find my service? Well, how many phone numbers do you remember now? You're not calling a number; you're calling a person (or sometimes an automated machine).

With Consul clients distributed throughout the network, tracking and monitoring your services, you don't need to know the addresses anymore. Don't worry about load balancers and firewalls, we'll get to that later. When the Consul service client is configured, it not only knows the address of its own machine but also recognizes what applications are running on the server through a set of intelligent health checks. These checks can be anything from a simple port check to a customized script to make sure the application is not just hearing but is actually listening (there is a difference). Once the server cluster is aware of the healthy application, and its location, then any application that can utilize the Domain Name Service (or DNS – you use this every time you use a web browser) can discover the application. There are other ways to share this information as well, but we'll discuss those later.

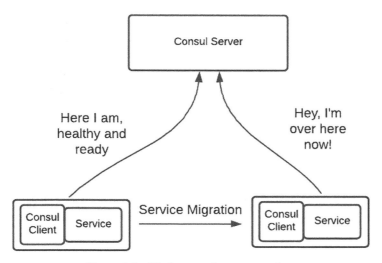

Figure 1.4 – Find your migratory services

As we've seen, the evolution of our network infrastructure has produced some amazing benefits when it comes to the management and the deployment velocity of the infrastructure. However, it has also introduced some complexity and perhaps some unwelcome variability, but with Consul service discovery, we're able to embrace those changes without trying to figure out who moved my cheese. So now that we know where our services are, what do we do next?

Something is meshy around here

If you've read any technical articles within the last couple of years, especially within the discussions of microservices or Kubernetes, you'll have heard the term **service mesh** quite pervasively. Personally, whenever I think of a mesh, it's rarely a clean concept of machines communicating with each other. Usually, I think of the myriad of tunnels that have been buried underneath Boston, but that's another story. From a network perspective, however, a service mesh provides a way to securely connect your various services without requiring additional functionality within the applications themselves. But it still sounds kind of meshy, so why would I want to deal with this?

In the previous section, we talked about service discovery and how easy it is to determine where different applications and services reside, along with their health status. With that information, a bad actor can target the service for a variety of attacks and even service *spoofing*. For example, if I figure out that a particular application is connecting to a service, some sort of data store or provider, it would be possible to create a rogue service that the application would unknowingly connect to and potentially divulge sensitive information. Our ability to secure our applications has certainly improved over the past several years, however, the problem of imposters, cowans, and eavesdroppers has persisted, if not gotten worse, within a dynamic cloud world. And when this happens, it usually isn't good.

The Consul service mesh provides a networking model that ensures that all communication from service to service is not only encrypted but there is validation that the sender and receiver are actually who they claim to be. A great analogy for these communication systems is the postal system.

Let's step through the process of sending a package, something important that we want to secure and verify that it made it to the final destination:

1. First, we're going to wrap the item we want to send in a box or envelope of some kind. On that package, we're going to place a label. Typically, that label will include the name of the person the package is destined for and their address. Note that this is a fixed address.

2. We're going to hand that package to a postal worker…for argument's sake, we'll just call them *Proxy*. Proxy is going to make sure we are who we say we are, and because this package is important, we're going to tell Proxy that we need to make sure it goes to the right person.

3. Proxy is going to potentially slap another label on the package and start moving the package through space and time. Since Proxy is local to your area, they aren't going with the package, but there is trust that the package will be protected along its journey.

The package may see all sorts of interesting things during its travels, but eventually, it will arrive at the name and address you intended. Proxy's distantly related sibling, *Envoy*, will validate that the package arrives at the proper location, based on the attached label. If we want to be fancy, we can even receive confirmation (such as an ACK) that it was delivered.

This simple analogy describes the communication within a service mesh. There is a message that needs to be communicated, but the application itself has neither the resources nor the intelligence to get that message securely to its intended destination. This is especially the case when that intended destination can change as, unlike our home address, we've already determined that we may not have the details of the destination address. So, the application hands that message to a local proxy, within the same machine, to eliminate the possibility of external snooping. That proxy has been configured with a certificate for two reasons. First, to validate its own authenticity. Second, to encrypt the message so that only those that are within the *inner circle* can obtain it. When the message gets to the other side, the receiving proxy can validate the certificate of the transmitter and unencrypt the message. From there, the proxy hands the message to the receiving application, again, on the same machine to eliminate any external snooping.

Figure 1.5 – Secure message transmission via the Consul service mesh

As we've seen, adding a service mesh into the Consul architecture enables us to not only discover our services but we can also securely, with validation, connect to those services. However, we may be using identical certificates within our network in order to validate and encrypt messages. This practice is not uncommon, and therefore opens the possibility of a rogue application within the mesh talking to whatever else it wants to. Well, we can't have that, so through Consul, we can create what's called an *intention*. When you create an intention within Consul, you're telling the associated Consul client who it can, and more importantly cannot, receive messages from. This provides yet another level of security within the service mesh.

OK, that service mesh stuff is a bit confusing, but I get it. Now that I'm not dealing with IP addresses, I can now set up connections between my applications and services that not only secure the message but also validate the sender. I can even control the communication within my own network by creating rules regarding who can talk to whom. But there is still all of this other network stuff that slows me down. Even in the cloud, I'm hit with delays with the firewall, load balancer, and other communication devices. Why can't we automate those pieces? Well, of course you can!

Automate me!

So, we've established how great it is to be able to discover the services available on our network without cumbersome static addressing, and how to utilize that to ensure the secure delivery of messages among our services. Wouldn't it be great to share this dynamic information with other components in the network? Well, why would we want to do that? To understand how this functionality can improve the lives of so many people, let's jump into the way-back machine and once again recall what we used to do, and often still do, when making adjustments to the network:

1. Yeah! We have our application deployed into the network and I can communicate with it. However, I can't reach the service my application needs to hit. Better call the network team.

2. The network team told me that the problem must be on my end because the destination service is *alive* and responding to their monitoring tools.

3. After a few days of troubleshooting, we learn that there is a firewall in the path that is protecting that service from nefarious creatures.

4. So, next, we submit a ticket to the firewall team and request a rule be configured to allow traffic from, you guessed it, my IP address.

5. Of course, the firewall team is not only getting hammered with requests, but they know that one wrong move on the list of firewall rules and the number of tickets will be the least of their problems, so extreme care must be taken.

6. Eventually, the team is able to safely add and test the new rule, and we're off and running.

Now after all of that, let's hope that we didn't forget to include any important information in that ticket (such as an additional port). Not to mention that when the address changes (not if), we'll need to enter the process again. This process continues today, even with cloud automation, for a number of different reasons. In my personal experience, these changes are the ones that tend to create the most confusion and delay in any project, and in a dynamic world, the changes aren't slowing down.

Hopefully, you can see how the knowledge that Consul is aware of plays such a critical role in this scenario. As services are added, updated, and removed, the Consul server cluster is aware of the changes, thanks to the distributed agents. Now the main challenge is how to get that information from Consul out to my network equipment, and there are a few ways to solve that.

One of the most straightforward methods for network devices to learn about service changes is for the equipment to learn about those changes directly from Consul via its API. This functionality is already in use with the Consul integration with the F5 Big-IP load balancer using the F5 Application Services interface. Whenever Consul discovers network updates, the load balancer learns of these updates and can adjust the load balancing pools automatically. Although this integration does require the network components to develop functionality specifically for the Consul API, there is no Consul agent required on the load balancer itself.

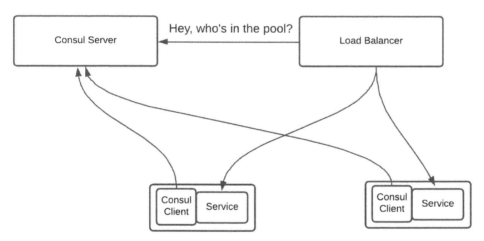

Figure 1.6 – Load balancer querying Consul for service updates

An alternative to this level of integration is to have Consul post updates to the service availability to a messaging service, such as ActiveMQ or RabbitMQ. Any messaging service, or other application, that can receive an HTTP message can receive service updates from Consul, without having to query it directly.

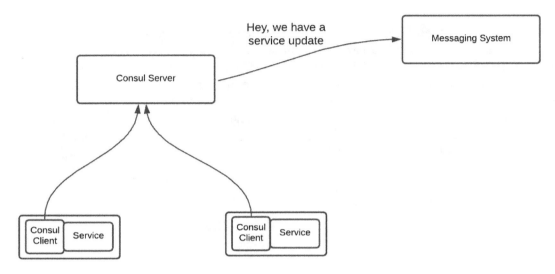

Figure 1.7 – Consul dynamically updating a message bus

A more common method of integration is utilizing Consul's template feature. Just as it sounds, you can create a configuration file template to match whatever device that you need to manage. Within that template, you specify areas where you need input from Consul, such as the IP address of a particular endpoint! If you remember, Consul makes intelligent decisions based on the data it receives. All of that data the Consul servers receive from the clients are pieces in the puzzle of an intelligent and dynamic network. A very simplified analogy would be setting up a payee for paying our bills. I presume this is something many of us do, certainly more common than writing checks. On our bank website, we establish a payee – some company or person that we pay money to every month. If we look at our monthly electricity bills, for example, the amount we pay varies based on the level of electricity my kids use that month. Why can't they ever learn to turn off the lights? So, we have our payee set up, and we know we are paying the electric company. That is our template. All we need is the data point of the amount due. Once the electric company (the client) informs us (the server) of the amount due, we can fill in the amount and let the electronic payment go. Hopefully, this will happen before the electricity is shut off!

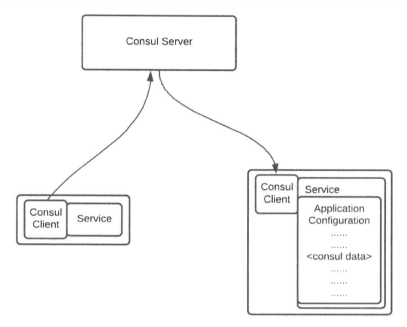

Figure 1.8 – Terrific templating

That template functionality, however, doesn't only apply to configuration files. HashiCorp does have several other products besides Consul, one of which is Terraform. If you aren't familiar with Terraform, it provides a standardized method to define system infrastructure as code for multiple cloud providers, but it also has providers for multiple applications. This includes firewalls, load balancers, monitoring systems, and so on. Now imagine coupling the dynamic aspect of Consul service discovery, with Terraform's infrastructure-as-code platform!

> **Important Note**
>
> The Consul synchronization functionality with Terraform, called Consul-Terraform-Sync, is in public beta at the time of writing and supports a limited set of partner modules for automation. With HashiCorp's development velocity, by the time you're reading this, the functionality is likely fully supported.

Summary

Congratulations, we've made it through *Chapter 1*! I hope it was as fun for you reading it, as it was for me writing it! To review, we started learning about the overall Consul architecture, and how the servers and clients work together to create a dynamic and intelligent system. Like real life, those clients are really the workhorses of the system, monitoring the services, and feeding the data to the servers. That data collected by the distributed clients is fed back to the Consul server cluster and made available for a variety of consumers. The need for dynamic discovery has become critical as we've moved to the cloud and automation. With the awareness of our services, we are able to ensure the validation of transmitters and receivers, and encrypt the information communicated among our components. All of this is possible without changing our applications! Of course, it would be a shame to keep all of that useful network information to ourselves, and we were able to see how Consul shares that information among various devices to simplify our lives and accelerate the overall application deployment flow.

I hope this first chapter was a great start to your Consul journey, and I thank you for putting up with my jokes. Of course, there are more to come! As we dig a little deeper in *Chapter 2, Architecture – How Does It Work?*, we're going to learn more details about how these servers and clients actually communicate. And if you like, we're going to build our first cluster!

2
Architecture – How Does It Work?

Now that we've had a managerial-level view of what Consul is and what it can do for us, we're going to take a deeper dive into the architecture of the Consul system and look at how it works. Throughout this chapter, we're going to be building up a Consul cluster on our own in order to better understand what information is distributed within the cluster, who is responsible for it, and how it is shared. The topics covered within this chapter include the following:

- Consul cluster operation – We've already seen in theory how Consul servers operate as the brain of the cluster, and how clients work to monitor the services. Now we'll put that theory into practice.

- Establishing consensus – We're going to take a look into how the Consul servers establish consensus to elect a leader and maintain the service catalog.

- Cluster communication protocols – There's a lot of activity happening among the cluster servers and the clients. We're going to dig a bit deeper into how all of that information is shared and updated.

- Geography – A lot of decisions are managed within Consul, and service location doesn't just mean the **Internet Protocol (IP)** address it is behind. We're going to see what other service criteria Consul can report.

As part of our educational journey, we will be creating our own Consul cluster within **Amazon Web Services (AWS)** using Terraform. If you've never used Terraform before, don't worry. We only need to perform a few basic instructions, and I've written all of the code for you! If you have used Terraform before, please don't laugh at my code; I never claimed to be a software developer.

Technical requirements

As you will be building your own cluster within this section, there are some requirements that we need to address. However, we'll be building everything in AWS, so all you really need is the ability to run Terraform from your command-line environment. I'll be using macOS, but Terraform can run in most desktop operating systems. Over the next few sections, we'll be setting up what you need within AWS and downloading the necessary HashiCorp tools. In order to work together as we proceed, you're going to want to create an AWS account if you don't already have one. We're going to be using AWS to create and manage the infrastructure throughout the process. An effort has been made to ensure that the costs of setting up this infrastructure are minimal, but there are a few things to be aware of, such as the following:

- Always perform a `terraform destroy` operation when you are done for the day. This does mean you'll have to rebuild as you continue working the next day, but what a great lesson on immutable infrastructure!

- Be very careful with your AWS credentials. Never share your **access key identifier**, and definitely don't share your **secret access key**. We're going to set them as environment variables so that they never show up in code and they should absolutely never show up in your **Version Control System (VCS)**!

Obtaining AWS credentials

To obtain your AWS credentials, log in to AWS at `https://console.aws.amazon.com/iam/home?#` and proceed to **Identity and Access Management (IAM)**.

On the left side of the page, you should see a **Users** section where you can utilize an existing user or add a new one. I would recommend creating a new user specifically for this project. You can see an overview of this page in the following screenshot:

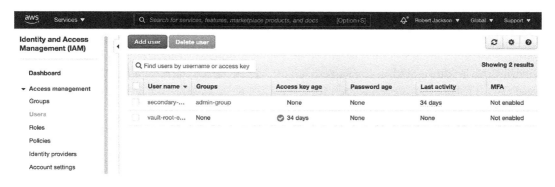

Figure 2.1 – AWS user management

Create a username that you're happy with and select **Programmatic access** for the access type, as illustrated in *Figure 2.2*. Hit the **Next** button and continue the process:

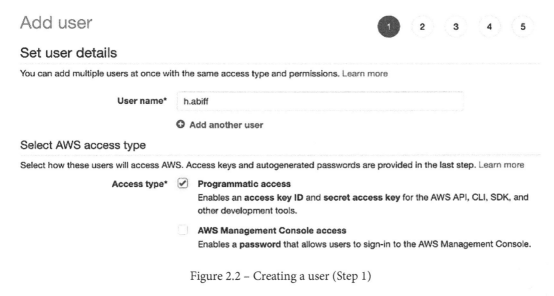

Figure 2.2 – Creating a user (Step 1)

For the next step, we're going to assign the appropriate permissions for the new user we're creating. We're going to assume that we don't have any groups or existing users within the system, so we're going to attach an existing policy as the permission set. Everything we're doing within the system is going to be utilizing the **Elastic Compute Cloud (EC2)** infrastructure, so search for `AmazonEC2FullAccess` and select that option for the user, as illustrated in the following screenshot:

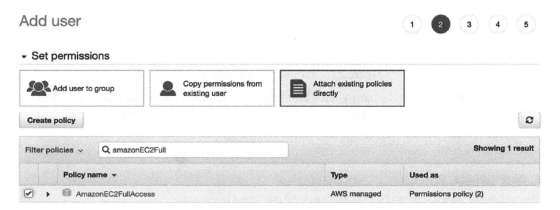

Figure 2.3 – Setting permissions for the user

Now that we have the permissions allocated, the next thing we're going to do is add a tag to the account so that it can be easily identified. Adding tags to our infrastructure or accounts is usually optional. However, it's always a good practice to help with identification, so throw in a tag (make it fun!), as illustrated in the following screenshot, and click the **Next** button again:

Add user

Add tags (optional)

IAM tags are key-value pairs you can add to your user. Tags can include user information, such as an email address, or can be descriptive, such as a job title. You can use the tags to organize, track, or control access for this user. Learn more

Key	Value (optional)	Remove
Purpose	Up and Running	✗
Add new key		

You can add 49 more tags.

Figure 2.4 – Adding tags to the user

Stick with us now…we're almost there! The next screen you'll see is a review of the new account information. If you've followed along like good girls and boys, you'll see something very similar to this:

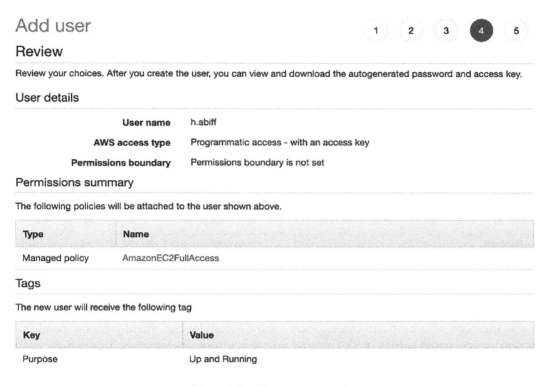

Figure 2.5 – User creation review

By now, I hope you've clicked your last **Next** button (well, at least for now) and can rejoice in the creation of a brand-new user! I don't know about you, but I always feel better when I see a big **Success** sign and a green checkbox! What's more important here, though, is the access credentials that have been created and assigned to your new user, as illustrated in the following screenshot:

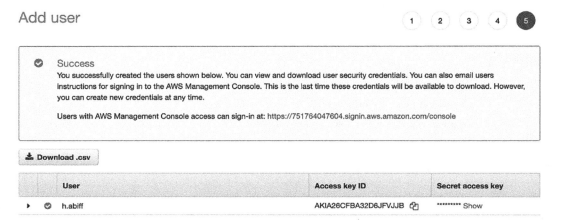

Figure 2.6 – User access credentials

OK—now that you've created your new user, you have a set of very specific and very important credentials that it would be very difficult, if not impossible, for any mortal being to memorize. To keep this information handy, I would recommend you download the **comma-separated values** (**CSV**) file and save it in a *very* safe location. This is not something you want to share with your favorite meme buddies, especially if they are anything like my friends!

Within that CSV file, you'll find your **User name**, **Access key ID**, and your **Secret access key**. Yeah—there's some other stuff in there, but that's not as useful. I've included a sample view of the CSV file here:

User name	Password	Access key ID	Secret access key	Console login link
h.abiff		AKIA26CFBA32D6JFVJJB	NYaSFhAzZqOq4bqSGJda/Y22Guc3pB2WCA8R9jVj	https://751764047604.signin.aws.amazon.com/console

Figure 2.7 – Access credentials CSV file contents

Yes—I understand I just exposed myself to anybody reading this book by sharing my credentials, something I said you should never do. Well, I just revoked those credentials, so not even the editor had a chance!

Now, we have some options of what we can do with this data, even if we aren't posting it to InstaFace. To utilize these credentials with Terraform, we're going to want to create environment variables with their content. We'll now see some examples, using the credentials shown in *Figure 2.7* as an example.

> **Note**
>
> These credentials can also be saved in files used directly by Terraform; however, that can sometimes lead to the credentials being shared or inadvertently distributed. Then, before you know it, you have a massive cloud computing bill and no Bitcoin to show for it!

One option is to manually copy and paste the credential values into `export` statements, like this:

```
$ export AWS_ACCESS_KEY_ID=AKIA26CFBA32D6JFVJJB
$ export AWS_SECRET_ACCESS_KEY=NYaSFhAzZqOq4bqSGJda/Y22Guc...
```

Alternatively, we can create environment variables using some Bash tools (assuming you're in the same directory as the downloaded file), as follows:

```
$ export AWS_ACCESS_KEY_ID=$(tail -n 1 new_user_credentials.csv
| awk -F, '{print $3}')
$ export AWS_SECRET_ACCESS_KEY =$(tail -n 1 new_user_
credentials.csv | awk -F, '{print $4}'
```

You'll know you did it correctly if you can validate that the values in the environment are the same as what is in the credentials file. You can do this by using the following command:

```
$ env | grep AWS
AWS_ACCESS_KEY_ID=AKIA26CFBA32D6JFVJJB
AWS_SECRET_ACCESS_KEY=NYaSFhAzZqOq4bqSGJda/Y22Guc3p...
```

Alternatively, if you are using Terraform Cloud, you can enter them as environment variables for the workspace, as shown in the following screenshot. However, keep in mind that we will primarily be working from the **command-line interface (CLI)**, so unless you are familiar with Terraform remote backends, it is advisable to simply use the Terraform CLI:

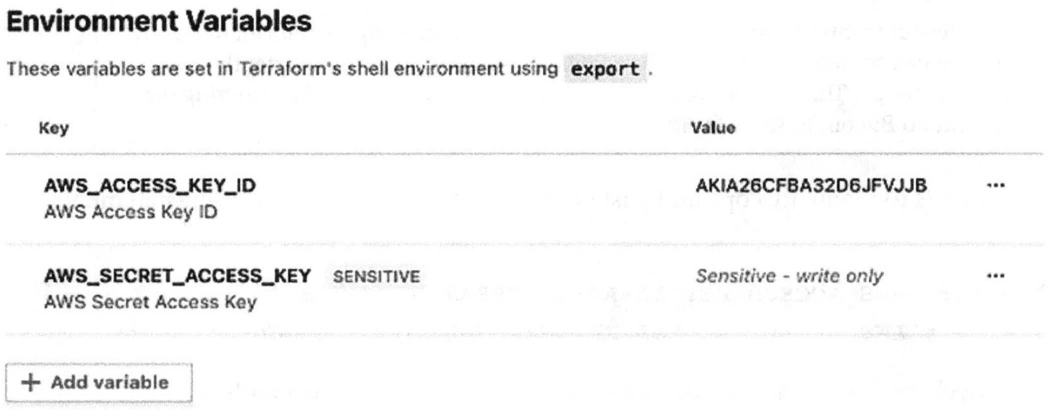

Environment Variables

These variables are set in Terraform's shell environment using `export`.

Key	Value	
AWS_ACCESS_KEY_ID AWS Access Key ID	AKIA26CFBA32D6JFVJJB	...
AWS_SECRET_ACCESS_KEY SENSITIVE AWS Secret Access Key	*Sensitive - write only*	...

+ Add variable

Figure 2.8 – Terraform cloud environment variables

Yeah! All of that work just to make sure that you have adequate credentials and that they are in the proper location.

Installing Terraform and Packer

We're almost there…can you feel it? Be sure to follow along closely here. We're going to be using Packer to create a base image for our projects and Terraform to create the infrastructure.

To install Terraform for our usage, start by downloading the appropriate ZIP file for your operating system from the Terraform download page, at `https://www.terraform.io/downloads.html`.

To install Packer for our usage, start by downloading the appropriate ZIP file for your operating system from the Packer download page, at `https://www.packer.io/downloads`.

For reference, the materials in this book were created with Terraform version 0.14.8 and Packer version 1.7.

Remember where the files are saved, and once the download is complete, unzip the files. You'll end up with a single `terraform` executable file and a single `packer` executable file that you need to save to a happy home and ensure that `home` is in your command-line path.

That's it! Now, you might be asking: *Rob, if it's that easy to download and use Terraform, why isn't everybody doing it?* That is a very good question, my dear reader…a *very* good question.

What a cluster!

If you've just set up your AWS account and installed Terraform, I can imagine you are ravenously awaiting the moment to fire up a cluster. Don't worry—as Robert Plant always used to say, *your time is gonna come*. First, we're going to dig a bit deeper into the mechanisms employed within the Consul cluster operation itself. I may have mentioned this before, but one of the core applications of Consul is to facilitate and enable a simpler and more secure communication mechanism among all of the services within the network. Consul would never be able to facilitate external communication if it didn't have a handle on its internal communication. So, let's look at the communication mechanisms and flows within the entire Consul cluster, both servers and clients.

Starting with the servers, what is the fundamental information they need to share and communicate? Here's what's required:

1. Leader election—After all, without a leader, there will be anarchy, and services will be burning in the streets.
2. Client registration—Servers obviously need to know when clients join the cluster.
3. Service status and location—The servers need to know the status of each service deployed within the cluster, as well as their location.

In order to elect a leader, the server nodes have to gain consensus. They do this using a protocol called—well—**consensus**. We're going to take a closer look at this protocol in an upcoming section, but essentially what happens is that when the server nodes spin up, there will be an election. If the server hasn't received notification that somebody stepped forward to be a leader, it will elect itself. If somebody else elected themselves first, then the server will vote for that initial leader. Eventually, all of the servers will gain *consensus* and there will be an undisputed leader. Again, we'll see how this works in more detail and discuss the potential issues in a later section.

The status of the deployed services and their location is something that isn't needed only by the server, but also by the clients distributed throughout the network. Now, this level of communication isn't anything new, right? The servers need to know what's happening with the clients and the client function in nearly any distributed system. However, as these systems scale, you end up with hundreds, thousands, or even tens of thousands of clients all trying to communicate with the servers. Eventually, the server leader is going to bang the gavel and insist on order in the court! With Consul, however, the servers and the clients all gossip with each other, using a protocol called—well—**gossip**. The imagination behind the names of these protocols is astounding. Utilizing the gossip protocol, clients share information about their own services and status with their neighboring clients, and the servers are all listening in. You can see an overview of this process in *Figure 2.9*:

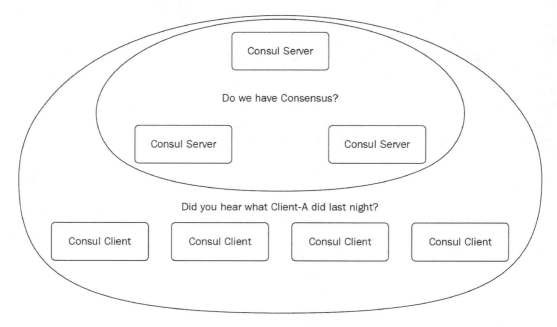

Figure 2.9 – Consensus among servers – everybody gossips

So, now that we have everybody talking, what are they all talking about? I want to know too! Don't leave me out! In a feeble attempt to maintain some sort of focus, we're going to first discuss the role of the servers and then move on to the clients.

From *Chapter 1, Consul Overview – Operation and Use Cases*, we've seen that the foundational use case for Consul is service discovery. The most critical aspect of this use case (and, therefore, all other use cases that follow) is the **service catalog**. The Consul servers are responsible for establishing and maintaining this service catalog, utilizing the information learned from the gossiping clients. Although the clients may have a plethora of information about the services, the servers only have to be concerned with the service status and the location…the information provided by Consul to those requesting the data.

The validity of the information within the service catalog is completely dependent upon the functionality performed by the clients. If the servers themselves had to maintain all of the health checks for every service on every client, the scalability of Consul just wouldn't be possible. I imagine it would be like Louis and Billy Ray on the stock market floor trading orange juice futures. For this reason, the distributed Consul clients perform all health checks for their associated services. Any failures of the service are reported by gossiping, and if the client itself fails, its neighbors recognize that it left the gossiping corner and alert the servers.

As stated before, the Consul servers learn this information by listening in on the gossip among the clients. However, what if rumors start to fly, as they often do amidst a world of gossip? We get confusion and uncertainty about what exactly is real within this world, and before we know it, we're sitting around a fire within a cloud of smoke discussing the meaning of life. In the technical world, this is called *entropy*, and it's a common problem among distributed systems. Consul addresses the problem of entropy by periodically synchronizing the information within its service catalog with the distributed agents. The rate at which the information is synchronized depends upon the size of the overall cluster. *Figure 2.10* lists the synchronization intervals for various Consul cluster sizes:

Cluster Size (Nodes)	Synchronization Interval
1-128	1 Minute
129-256	2 Minutes
257-512	3 Minutes
513-1024	4 Minutes
1025-2048	5 Minutes

Figure 2.10 – Cluster size versus time to synchronize

Hopefully, you are able to follow a pattern as the cluster size increases, correlated to the synchronization interval. This pattern continues. As you can see, even if the servers become discombobulated and are out of touch with the clients, it doesn't take long to get back in sync and correct any service catalog inaccuracies.

All right—now that we know essentially what happens within a massive cluster, let's take a deeper dive into the protocols used that help make all of this happen!

Rafting consensus

We've seen how the Consul servers elect their leader utilizing the **consensus** protocol, but what exactly does that mean, and is that all that they do? Well, of course they do more; otherwise, this would be a very short section. The consensus protocol that Consul utilizes is based on the Raft algorithm (`https://raft.github.io/raft.pdf`), which is based on Paxos (`https://en.wikipedia.org/wiki/Paxos_%28computer_science%29`). If you get the feeling that we are quickly heading down a rabbit hole, don't worry—we aren't going to keep digging until we hit water.

A closer look at leader election

As you will soon see, a Consul server is configured with a number of other servers, which it will Raft up with. This is called the `bootstrap_expect` value within the configuration. That number of servers is important to remember, as it will come up again. Of course, if a Consul server is going to expect a number of servers to Raft up with, it needs to know about them so that is also configured. In order to become a leader, a server needs to be elected by a majority of servers within the Raft—for example, if you are expecting five servers within your Raft, you'll need at least three to vote for you. This is referred to as a quorum and is a critical concept when it comes to the availability of a server cluster. Much like with the many meetings that we hold, if you don't have a quorum of members, you can't really make any decisions. In some meetings, even with a quorum, you can't make any decisions if you don't have anybody leading! If you aren't sure about how many servers are needed to attain a quorum, refer to the following equation:

(Number of Servers)/2 + 1 = Quorum

If the number of servers is the same as the quorum size, there is no protection for the cluster. Of course, you can't have a half of a server, so if you end up with a half, just round down. The table in *Figure 2.11* might also help:

Number of Servers	Quorum	Failure Tolerance
1	1	0
2	2	0
3	2	1
4	3	1
5	3	2
6	3	2
7	4	3

Figure 2.11 – Never fail for quorum

So, let's look at how that election is performed. For a really cool animated view of the process, the following website is fantastic: `http://thesecretlivesofdata.com/raft/`.

Walking through the election, we're going to look at three servers: **Bart**, **Lisa**, and **Maggie**. **Bart** is the first server to recognize consciousness, so he submits himself as a *candidate* and sends a request to **Lisa** and **Maggie** to become the *leader*. However, before **Lisa** heard **Bart**'s request, she sent her own request for candidacy with a leader request to **Bart** and **Maggie**, because we all know **Lisa** is smarter.

You can see an overview of the process in *Figure 2.12*:

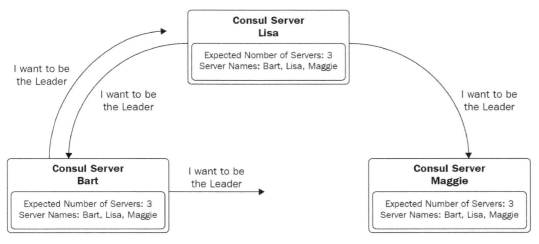

Figure 2.12 – Lisa's nomination for election

Maggie hears **Lisa**'s request for leadership and responds happily. By this time, **Bart**'s request has made it to both **Lisa** and **Maggie**, but alas, he is too late. **Maggie** has already voted for **Lisa** (I mean, seriously: why would she vote for **Bart** anyway?).

You can see an overview of the process in *Figure 2.13*:

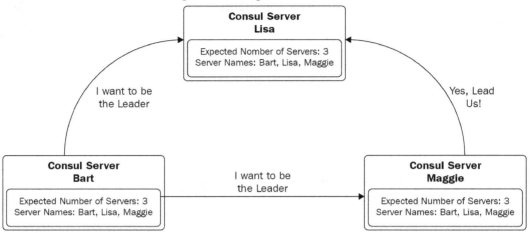

Figure 2.13 – Maggie submits her vote

At this point in time, **Lisa** and **Maggie** both agree that **Lisa** should be the leader. Although **Bart** might protest, he is in the minority as **Lisa** and **Maggie**'s agreement satisfies the quorum. Now, **Lisa** is the *leader*, and **Maggie** and **Bart** have become *followers*. The group of **Bart**, **Lisa**, and **Maggie** now have a consensus and create their own *peer set* of Consul servers.

You can see an overview of the process in *Figure 2.14*:

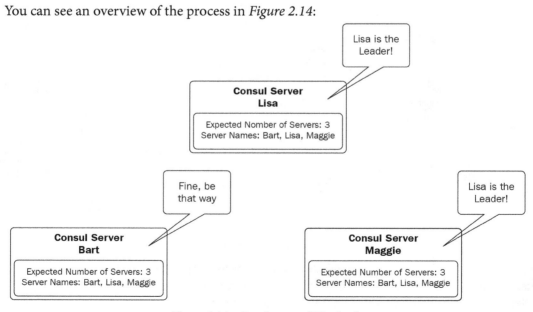

Figure 2.14 – Bart loses and Lisa leads

Wonderful—we now have a leader for the cluster! However, the election of a server leader is only one piece of what Consul actually utilizes the consensus protocol for—after all, what is a leader if it doesn't share the information it learns?

> **Note**
>
> Although this process can move quite quickly with only three servers, it only accommodates the failure of a single Consul server. You may be thinking: *Wow, I'll just throw in, like, 12 servers and I'll be good!* There is a trade-off—as the number of servers in the Raft increases, so does the time it takes to achieve a quorum. This not only impacts leader elections but also data consistency. Usually, it is recommended to keep the peer set of servers between five and seven.

Data sharing within the server peer set

Now that **Lisa** has become the undisputed leader, she is responsible for maintaining the information shared with **Bart** and **Maggie**. This information is recorded and shared via log entries written by **Lisa**. These log entries are what contain the state of our service catalog and leader election process, and essentially anything that the server peer set needs to know in order to function. It's time to start peeling that onion, but I do hope it doesn't make you cry.

Within the peer set, any server can receive information from the clients and share that information with the leader. This information can contain a change to the service status on the direct node or even updates to other nodes within the cluster. The leader is the only one that updates the log entries; however, just because the leader, **Lisa**, knows about some adjustment to the service catalog or other components, it doesn't mean that **Maggie** or **Bart** have the same information. This is where our quorum comes back into play. Just as with *groupthink*, a minimum number of servers within the peer set need to agree on what is the reality at that point in time. Therefore, as **Lisa** writes an update to her log file, she sends that update to both **Bart** and **Maggie**. Until either **Bart** or **Maggie** confirm that they've applied the corresponding log entry and the log has been properly *replicated*, the entry is not considered to be *committed*. However, once **Lisa** gets confirmation and at least one other member of the server peer set (remember—there are three members, so we need two to achieve a quorum), we can consider the log entry to be committed and we have a new view of the reality.

All of these exchanges utilize Remote Procedure Calls (RPC) and consensus, as displayed in *Figure 2.15*.

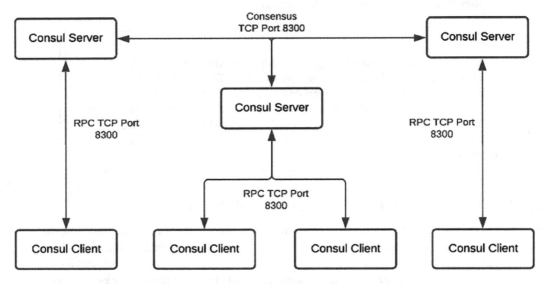

Figure 2.15 – Consensus and RPC connections within the peer set

I know what you may be thinking: *OK, Rob, but what if the logs grow too large?* This is a valid concern and is managed via a common **finite-state machine** (FSM) within the server peer set. As logs are replicated and committed among the servers, the FSM is updated with the current status. When queries are made to the Consul server peer set, the leader responds to that query based on the status of the FSM. Consul is structured in such a way that any replay of a set of logs will result in an identical state. Wow—wouldn't it be nice if humans were that predictable? This creation and maintenance of the state, however, enables Consul to periodically snapshot the log entries in order to manage the replicated log size. This snapshot is performed without any interruption to Consul services and ensures long-term happiness among the peer set.

Gossiping serfs

Now that we understand a bit more about how a server peer set is established and operates, let's take a deeper look into what everybody is gossiping about. Again, if we start unpacking where the gossip protocol comes from, it utilizes the **Serf** (https://www.serf.io/) library, which is based on the **Scalable Weakly-consistent Infection-style Process Group Membership** protocol, which has thankfully been shortened to **SWIM** (http://www.cs.cornell.edu/info/projects/spinglass/public_pdfs/swim.pdf). We've jumped off of the Raft, and we've gone swimming!

Have a look at the following diagram:

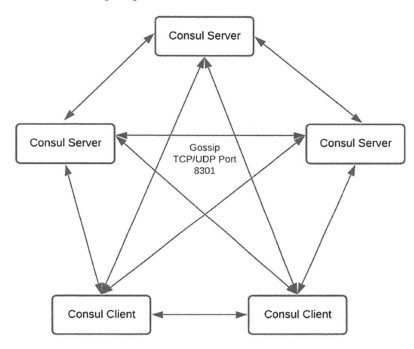

Figure 2.16 – Gossip connections

All of the members of the Consul cluster, the servers and clients, are swimming in the same pool. This enables the distributed discovery of all members of the pool, their status, and—of course—the status of any services registered with the Consul clients. Although gossip is how all of the agents learn about changes to cluster members as well as the services, all information is still sent from the agents to the server nodes via direct RPCs. Remember, any server within the peer set can receive an update and forward this to the leader for quorum commitment and an update to the state machine.

Changes in latitudes

We've already seen how Consul shares information within a cluster regarding leadership election, changes in service availability, and changes in node availability. In addition to these items, Consul is always aware of the geographic location of each node relative to others within the cluster. This can help with intelligent decisions such as locating the nearest available service instance to the requesting entity. The discovery mechanism utilizes network tomography offered by the Serf library. Yup—not only do they all gossip about each other's state, but they also know each other's location! The singularity is upon us!

Setting up our first Consul cluster

The hour has arrived! At the beginning of this chapter, there were instructions for creating credentials for an AWS account, as well as for downloading Terraform. Now, you finally get to put all of that work to good use!

> **Note**
>
> This book covers aspects of Consul for both server and client operations. HashiCorp also offers a managed Consul service on **HashiCorp Cloud Platform** (**HCP**), which may be desirable when deploying within your own production environment. However, in order to understand the server operation, we'll be operating on our own. For more information about HashiCorp's hosted Consul service, visit `https://portal.cloud.hashicorp.com/`.

All of the code for this section is in the following GitHub repository:

`https://github.com/PacktPublishing/Simplifying-Service-Management-with-Consul`

You can either fork the repository or just download it and play along. We're going to be building our system for each chapter with code in that repository, with the relevant code in the appropriate chapter folder.

The following sections include a review of the Packer and Terraform code that we'll be using to create our Consul cluster. It isn't necessary to understand every aspect of the code, but I wanted to include explanations for completeness. As you proceed, however, take careful note of the commands indicated by the $ symbol.

Building the image

We're going to start by building our base image. In order to keep us all on the same page—quite literally—regarding the image being used and the operation of Consul, we're going to be building an image together using Packer. But don't worry—all of the code you need is in the repository. Within the `ch2` folder, you'll find another folder called `Image-Creation`. I know—very inventive. Here's the code you'll need to get to this folder:

```
$ cd $github_root/Consul-Up-and-Running/ch2/Image-Creation
```

There are a few key items in here to be aware of. Your variables for the build are all defined in `variables.pkr.hcl`, as illustrated in the following code snippet:

```
variable "owner" {
  type        = string
```

```
  description = "Owner tag to which the artifacts belong"
  default     = "packt-consul"
}
variable "consul_version" {
  type = string
  description = "Three digit Consul version to work with"
  default = "1.9.0"
}
variable "aws_region" {
  type        = string
  description = "AWS Region for image"
  default     = "us-east-2"
}
variable "aws_instance_type" {
  type        = string
  description = "Instance Type for Image"
  default     = "t2.small"
}
```

For most of these, the defaults will be fine. However, if you want to change any of the defaults, make that change in `variables.pkrvars.hcl`. For example, look at the following snippet:

```
owner = "rjackson"
```

Yup—customizing your variables is just that easy! The Packer file that defines what we are building is called `AWS_linux_image.pkr.hcl`. Let's take a quick look at that file, just to see what we are building.

Lines *1* through *19* define the source image that we're going to be using to build our Consul image. We have variables set for the name and region, utilizing the aforementioned variable files. We're also filtering through AWS to find the appropriate Ubuntu image, as illustrated in the following code snippet:

AWS_linux_image.pkr.hcl

```
source "amazon-ebs" "ubuntu-image" {
  ami_name = "${var.owner}-consul-{{timestamp}}"
  region = "${var.aws_region}"
```

```
instance_type = var.aws_instance_type
tags = {
  Name = "${var.owner}-consul"
}
source_ami_filter {
   filters = {
     virtualization-type = "hvm"
     name = "ubuntu/images/*ubuntu-bionic-18.04-amd64-
server-*"
     root-device-type = "ebs"
   }
   owners = ["099720109477"]
   most_recent = true
}
communicator = "ssh"
ssh_username = "ubuntu"
}
```

Lines *21-58* actually build the image, including all the provisioning that we need. Essentially, we are taking that initial image we searched for and packing what we need into it—hence the name *Packer*. As part of this process, we are moving a couple of files over for Consul to utilize: one `service` file defining how Consul is to be managed within the Ubuntu environment, as well as a `common` configuration file that will be applied to all Consul nodes. We're also performing some inline provisioning, instructing Packer to `ssh` to the nodes and execute a set of commands as part of the setup. The code is illustrated in the following snippet:

```
build {
  sources = [
    "source.amazon-ebs.ubuntu-image"
  ]
  provisioner "file" {
    source      = "../files/consul.service"
    destination = "/tmp/consul.service"
  }
  provisioner "file" {
    source      = "../files/consul-common.hcl"
    destination = "/tmp/consul-common.hcl"
```

```
  }
  provisioner "shell" {
    inline = [
      "sleep 30",
      "sudo apt-get update",
      "sudo apt install unzip -y",
      "sudo apt install default-jre -y",
      "curl -k -O \"https://releases.hashicorp.com/
consul/${var.consul_version}/consul_${var.consul_version}_
linux_amd64.zip\"",
      "unzip consul_${var.consul_version}_linux_amd64.zip",
      "sudo mv consul /usr/local/bin"
    ]
  }
  provisioner "shell"{
    inline = [
      "sudo /usr/local/bin/consul -autocomplete-install",
      "sudo useradd --system --home /etc/consul/consul.d
--shell /bin/false consul",
      "sudo mkdir /etc/consul /etc/consul/consul.d /etc/consul/
logs /var/lib/consul/ /var/run/consul/",
      "sudo chown -R consul:consul /etc/consul /var/lib/consul/
/var/run/consul/",
      "sudo chmod -R a+r /etc/consul/logs/",
      "sudo mv /tmp/consul.service /etc/systemd/system/consul.
service",
      "sudo mv /tmp/consul-common.hcl /etc/consul/consul.d/
consul-common.hcl"

    ]
  }
}
```

The `service` file that defines how Consul is to be run as a service within Ubuntu is not unlike many others; however, there are a few lines that you should take a look at. Pay particular attention to the setting that identifies the Consul configuration directory (`config-dir`) and the setting that identifies where the Consul data (Raft log files and state information) will reside (`data-dir`). Any configuration files within that configuration directory will be read by Consul and applied to the configuration. This helps prevent one massive configuration file; however, multiple files can make things a bit more confusing when analyzing the configuration. Have a look at the following code snippet:

consul.service

```
[Unit]
Description=Consul server agent
Requires=network-online.target
After=network-online.target

[Service]
User=consul
Group=consul
PIDFile=/var/run/consul/consul.pid
PermissionsStartOnly=true
ExecStartPre=-/bin/mkdir -p /var/run/consul
ExecStartPre=/bin/chown -R consul:consul /var/run/consul
ExecStart=/usr/local/bin/consul agent \
    -config-dir=/etc/consul/consul.d/ \
    -data-dir=/var/lib/consul/ \
    -pid-file=/var/run/consul/consul.pid
ExecReload=/bin/kill -HUP $MAINPID
KillMode=process
KillSignal=SIGTERM
Restart=on-failure
RestartSec=42s

[Install]
WantedBy=multi-user.target
```

One of those configuration files includes the common settings that we want across all Consul nodes. For our purposes, we're only going to set this to the log settings. We're going to set these to DEBUG so that we can poke around and learn more about what Consul is doing! Have a look at the following code snippet:

consul-common.hcl

```
log_file = "/etc/consul/logs/"
log_level = "DEBUG"
```

At this point, you might want to double-check to make sure your AWS environment variables are set. You can do this with the following command:

```
$ env |grep AWS
AWS_ACCESS_KEY_ID=AFAIKJENNY
AWS_SECRET_ACCESS_KEY=8675309...
```

Now, we're ready to build! Execute the simple Packer command shown in the following code snippet and take a few minutes to congratulate yourself for packing an image! Celebrate with a beverage of choice and relax, as the process will take about 5 minutes:

```
$ packer build -var-file="variables.pkrvars.hcl" .
```

Reviewing the output, it will look like somebody threw Christmas lights all over your terminal. The Packer output is all red and in various shades of green, just to make things festive. You'll see files getting moved over, as well as the operating system updating, just to make sure we've got all of the latest packages for Ubuntu. You'll also see some lines about provisioning with a shell script. Packer creates shell scripts based on those shell provisioning blocks we saw within the Packer build definition.

At the end of the process, you'll see the location and the **Amazon Machine Image (AMI)** ID for your new image! This is illustrated in the following snippet. Now, you can build upon your Consul knowledge *in your own image*:

```
==> Wait completed after 5 minutes 52 seconds
==> Builds finished. The artifacts of successful builds are:
--> amazon-ebs.ubuntu-image: AMIs were created:
us-east-2: ami-079a52bd81f91b201
```

Again, there is no need to remember that image ID, as in our next step, Terraform will find it for us!

Building the Consul cluster

Start by changing directories to the root directory of ch2. This is where we will find the Terraform scripts for building our cluster. You can do this by running the following command:

```
$ cd $github_root/Consul-Up-and-Running/ch2/
```

Much like when building our Packer image, we have a few files in this directory to be aware of. Don't be concerned about the files directory just yet, as we'll get into that as we evaluate the Terraform script for building our cluster.

If you take a look at our variables.tf file, this is where the variables that Terraform requires are defined. I always get confused between variables being defined and where they are set. Anyhow, for each of the five variables, there is a description (always a good idea) and a reasonable default (well, reasonable from my perspective). Have a look at the following code snippet:

variables.tf

```
variable "aws_region" {
  type        = string
  default     = "us-east-1"
  description = "Region for AWS Components"
}
variable "owner" {
  type    = string
  default = "consul-demo"
}
variable "consul_region" {
  type        = string
  description = "Region of consul server (not AWS Region)"
  default     = "global"
}
variable "vpc_cidr" {
  type        = string
  description = "CIDR of the VPC"
  default     = "192.168.100.0/24"
}
variable "instance_type" {
```

```
type        = string
description = "machine instance type"
default     = "t3.micro"
}
```

If you would rather use your own values for the necessary variables, don't worry—I won't be offended. These variables are set within the `terraform.tfvars` file. Have a look at the following code snippet:

Terraform.tfvars

```
owner         = "rjackson"
aws_region    = "us-east-2"
instance_type = "t2.small"
```

OK—now that we've set our variables to our liking, it's time to fire up our infrastructure! If you've never used Terraform before, don't worry—it's quite easy, and all of the code has been built. Again, I never claimed to be a software developer, so please don't laugh at my code! First, we're going to kick off our infrastructure build and then dissect what we built. To get started, Terraform needs to download any provider-specific data necessary. This is performed via the following command:

```
$ terraform init
```

Of course, it's never good to build something without examining what you build, so next, we're going to plan our build. All of the output from this command should be green plus signs (+), identifying which infrastructure is going to be created. Note that many of these items are *known after apply* operation as they require assignment from AWS, but it might be fun to peruse what you are about to build. If not, don't worry—I'll review the details after we kick off the build.

Now, run the following command:

```
$ terraform plan
```

Next, we're going to apply our infrastructure build. This is done using—you guessed it: `terraform apply`. This kicks off another plan and shows you what you are about to build, but since we've already done that, let's live dangerously and *auto-approve*. Have a look at the following code snippet:

```
$ terraform apply --auto-approve
```

> **Note**
>
> The author can't be held responsible for any costs you incur during this practice. Please be sure to destroy your infrastructure when you quit for the day or night. If you want to set up an m3.xlarge system and leave it running for days, that's up to you!

All right—let's take a look at what we are building. All of the following code blocks (in this section) are within the main.tf file. It is Terraform best practice to have all pieces of the code in one file, so that's what we're going to do here!

This first section is fairly simple and simply sets up the Terraform providers that we'll be using. A Terraform provider is kind of like a module that provides functionality specific to the provider—for example, we're installing the AWS provider, enabling us to manage resources in AWS. We're specifying the version of that provider, and the region that we'll be working in. The provider block is also where naughty people would insert their AWS credentials, but we're *never* going to do that—WILL WE??? Finally, we're using an AWS data source to find the image that we just built using Packer. Here's the code you'll need:

```
terraform {
  required_providers {
    aws = {
      source  = "hashicorp/aws"
      version = "~> 2.5"
    }
  }
}
provider "aws" {
  region  = var.aws_region
}
data "aws_ami" "an_image" {
  most_recent     = true
  owners          = ["self"]
  filter {
    name   = "name"
    values = ["${var.owner}-consul-*"]
  }
}
```

This next section creates a private key for us to use in order to ssh into our machines. An organization would typically maintain its own private keys for machine access, but these really shouldn't be floating around. A more secure approach would be to utilize the HashiCorp Vault **Secure Shell** (**SSH**) helper for one-time-use SSH keys, but that's another story. The key will be built within our Consul machines, but we'll also have a local **Privacy Enhanced Mail** (**PEM**) file ending in -consul-key.pem. The code you'll need is illustrated in the following snippet:

```
resource "tls_private_key" "my_private_key" {
 algorithm = "RSA"
 rsa_bits = 4096
}
resource "local_file" "private_key" {
 content = tls_private_key.my_private_key.private_key_pem
 filename = "${var.owner}-consul-key.pem"
 file_permission = 0400
}
resource "aws_key_pair" "consul_key" {
 key_name = "${var.owner}-consul-key"
 public_key = tls_private_key.my_private_key.public_key_openssh
}
```

This section provides the vast majority of the networking information necessary for our system. As we aren't reading a book about the intricacies of AWS networking, you'll have to just trust me that all of the code in the following snippet is cool:

```
resource aws_vpc "consul-demo" {
  cidr_block          = var.vpc_cidr
  enable_dns_hostnames = true
  tags = {
    Name = "${var.owner}-vpc"
  }
}
resource aws_subnet "consul-demo" {
  vpc_id     = aws_vpc.consul-demo.id
  cidr_block = var.vpc_cidr
  tags = {
    name = "${var.owner}-subnet"
```

```
    }
}
resource aws_internet_gateway "consul-demo" {
  vpc_id = aws_vpc.consul-demo.id

  tags = {
    Name = "${var.owner}-internet-gateway"
  }
}
resource aws_route_table "consul-demo" {
  vpc_id = aws_vpc.consul-demo.id
  route {
    cidr_block = "0.0.0.0/0"
    gateway_id = aws_internet_gateway.consul-demo.id
  }
}
resource aws_route_table_association "consul-demo" {
  subnet_id       = aws_subnet.consul-demo.id
  route_table_id = aws_route_table.consul-demo.id
}
```

OK—this part of the networking is not something to gloss over. AWS security groups identify which ports are open between which devices. Here is a list of the defined ports and their purpose:

- Port 22—SSH access.

- Ports 8300-8302—These cover all consensus, RPC, and gossip communication between the nodes.

- Port 8500—This is used for the Consul web **user interface** (UI).

- Port 8600—Consul utilizes port 8600 for **Domain Name System** (DNS) queries as part of the Service Discovery function.

These ports have all been addressed and configured within the security group configuration. Here's the code illustrating this:

```
resource aws_security_group "consul-demo" {
  name    = "${var.owner}-security-group"
  vpc_id = aws_vpc.consul-demo.id
```

```
  ingress {
    from_port  = 22
    to_port    = 22
    protocol   = "tcp"
    cidr_blocks = ["0.0.0.0/0"]
  }
  ingress {
    from_port  = 8300
    to_port    = 8302
    protocol   = "tcp"
    cidr_blocks = ["0.0.0.0/0"]
  }
  ingress {
    from_port  = 8500
    to_port    = 8500
    protocol   = "tcp"
    cidr_blocks = ["0.0.0.0/0"]
  }
  ingress {
    from_port  = 8600
    to_port    = 8600
    protocol   = "tcp"
    cidr_blocks = ["0.0.0.0/0"]
  }
  egress {
    from_port      = 0
    to_port        = 0
    protocol       = "-1"
    cidr_blocks    = ["0.0.0.0/0"]
    prefix_list_ids = []
  }
  tags = {
    Name = "${var.owner}-security-group"
  }
}
```

All right—now, the good stuff! Within this section, we're creating three Consul servers, utilizing the AMI that we found as the data source (the AMI that we built using Packer). We're using the SSH key that we created as part of the Terraform script and have allocated a public IP address so that we can reach the server. Toward the bottom, we are associating tags with the servers so that we can easily find them. Do yourself and your IT operations group a favor: *never* deploy infrastructure without associating logical tags! One of those tags uniquely identifies the instance as well, which we can use to uniquely identify the server. Directly above the tags, there is a `user_data` field. This is pretty important as it calls a template file, `server_template.tpl`, to configure the Consul servers. We're also passing a `server_name_tag` property to the template file to help identify the servers. The code is illustrated in the following snippet:

```
resource aws_instance "consul-server" {
  count                         = 3
  ami                           = data.aws_ami.an_image.id
  instance_type                 = var.instance_type
  key_name                      = aws_key_pair.consul_key.key_
name
  associate_public_ip_address = true
  subnet_id                     = aws_subnet.consul-demo.id
  vpc_security_group_ids        = [aws_security_group.consul-
demo.id]
  iam_instance_profile          = aws_iam_instance_profile.
instance_profile.name
  user_data = templatefile("files/server_template.tpl", {
server_name_tag = "${var.owner}-consul-server-instance"})
  tags = {
    Name    = "${var.owner}-consul-server-instance"
    Owner = var.owner
    Instance = "${var.owner}-consul-server-instance-${count.
index}"
  }
}
```

In the next code snippet, we're building our Consul clients—four of them to be exact, representing the core elements of earth, air, fire, and water. Many of these parameters are identical to what we used for the servers, but notice that there is no `user_data` section. For variety, we're using a Terraform provisioner to configure the clients, although in a production environment, configuration managers such as Ansible would be preferred:

```
resource aws_instance "consul-client" {
  count                        = 4
  ami                          = data.aws_ami.an_image.id
  instance_type                = var.instance_type
  key_name                     = aws_key_pair.consul_key.key_
name
  associate_public_ip_address  = true
  subnet_id                    = aws_subnet.consul-demo.id
  vpc_security_group_ids       = [aws_security_group.consul-
demo.id]
  iam_instance_profile         = aws_iam_instance_profile.
instance_profile.name
  tags = {
    Name  = "${var.owner}-consul-client-instance-${count.
index}"
    Owner = var.owner
  }
}
```

It should be noted that remote provisioners aren't preferred within the world of Terraform; however, I find it very useful on occasion to customize the resources that we require. As part of this provisioning, we're iterating over each of our four clients and performing a few functions, as follows:

- Transferring a service definition file called `httpd.json` to the client. We'll discuss this more later as we start playing with the cluster.

- Creating a `consul-client.hcl` configuration file. I want to dig into this file a bit more:

  - We're advertising the public IP address of the client. This tells the Consul servers, and any devices requesting access to any services on this client, which address it can be reached on. We're choosing the public address of the client so that it can be reached by applications outside of the internal network.

- We're telling the agent not to run as a server, because—well—we're a client.

- We're enabling script checks. This allows custom scripts to be used to check for service health and is not enabled by default.

- We're telling Consul to bind its service to the private address of the machine. Note the difference between the bind address and the advertising address. The private address is known by the machine itself and identifies the external interface. The public address we're advertising isn't even known by the client machine itself and is managed by AWS.

- We're telling the clients where the servers are. Remember that a client needs to register with the servers in order to communicate the status of itself and all of its managed services.

- We're identifying the client address to use when communicating with the Consul server. Because the server and client reside within the same AWS virtual network, we can use the private IP address.

- Moving the client configuration and the service definition files to the configuration directory (/etc/consul/consul.d/).

- Starting a simple **HyperText Transfer Protocol (HTTP)** server using Python. After all, why go through the hassle of creating a Consul cluster without having any services running on it?!

- Next, we're using a provisioner to remotely start the Consul service on the client machine.

- This last block tells Terraform how it's going to connect to the client machine. Note that we're using the contents of that SSH key we created previously.

Notice that although the configuration for the clients is all in this Terraform file, the application configuration for the servers is not. This is because we used that user_data feature within Terraform, and we'll be looking at the template file very shortly. Within Terraform, when we utilize a provisioner for scripting or moving files, we declare a null_ resource property, as follows:

```
resource null_resource "provisioning-clients" {
  for_each = { for client in aws_instance.consul-client :
client.tags.Name => client }
  provisioner "file" {
    source      = "files/httpd.json"
    destination = "/tmp/httpd.json"
  }
}
```

```
  provisioner "remote-exec" {
    inline = [
      "sudo cat << EOF > /tmp/consul-client.hcl",
      "advertise_addr = \"${each.value.public_ip}\"",
      "server = false",
      "enable_script_checks = true",
      "bind_addr = \"${each.value.private_ip}\"",
      "retry_join = [\"${aws_instance.consul-server[0].public_
ip}\",\"${aws_instance.consul-server[1].public_ip}\",\"${aws_
instance.consul-server[2].public_ip}\"]",
      "client_addr = \"${each.value.private_ip}\"",
      "EOF",
      "sudo mv /tmp/consul-client.hcl /etc/consul/consul.d/
consul-client.hcl",
      "sudo mv /tmp/httpd.json /etc/consul/consul.d/httpd.
json",
      "nohup python3 -m http.server 8080 &",
      "sleep 60"
    ]
  }
  provisioner "remote-exec" {
    inline = [
      "sudo systemctl start consul",
    ]
  }
  connection {
    type        = "ssh"
    user        = "ubuntu"
    private_key = local_file.private_key.content
    host        = each.value.public_ip
  }
}
```

Next, we have the instance identity information and associated roles. This doesn't have any direct association with Consul but is required to enable the communication paths we need within AWS. The code is illustrated in the following snippet:

```
resource "aws_iam_instance_profile" "instance_profile" {
  name_prefix = var.owner
  role        = aws_iam_role.instance_role.name
}
resource "aws_iam_role" "instance_role" {
  name_prefix         = var.owner
  assume_role_policy = data.aws_iam_policy_document.instance_role.json
}
data "aws_iam_policy_document" "instance_role" {
  statement {
    effect = "Allow"
    actions = ["sts:AssumeRole"]


    principals {
      type        = "Service"
      identifiers = ["ec2.amazonaws.com"]
    }
  }
}
resource "aws_iam_role_policy" "metadata_access" {
  name   = "metadata_access"
  role   = aws_iam_role.instance_role.id
  policy = data.aws_iam_policy_document.metadata_access.json
}
data "aws_iam_policy_document" "metadata_access" {
  statement {
    effect = "Allow"
    actions = [
      "ec2:DescribeInstances",
    ]
    resources = ["*"]
```

```
    }
}
```

OK—now that we've gone through the full Terraform file, let's take a closer look at that server template file. Via the template, we're creating a `consul-server.hcl` server configuration file and populating it with all of the configuration necessary, as follows:

```
#!/usr/bin/env bash

cat << EOF > /etc/consul/consul.d/consul-server.hcl
server = true
bootstrap_expect = 3
retry_join = ["provider=aws tag_key=Name tag_value=${server_
name_tag}"]
bind_addr = "$(curl -s http://169.254.169.254/latest/meta-data/
local-ipv4)"
advertise_addr = "$(curl -s http://169.254.169.254/latest/meta-
data/public-ipv4)"
client_addr = "$(curl -s http://169.254.169.254/latest/meta-
data/local-ipv4)"
ui = true
EOF

# Starting consul services
sudo systemctl start consul
```

Following along with that template file, let's take a look at exactly what that file is doing:

1. Indicating that the Consul agent should operate as a server.

2. Next, we're identifying the expected number of servers that we want to see in our cluster. Remember that we mentioned this before when discussing consensus and quorums. We're creating three Consul servers, which means our quorum is… (I'll give you a hint—it's more than one, and less than three).

3. OK—this part is pretty cool. If you remember, in the Terraform script, we had that `instance` tag that identifies the individual servers. This `retry_join` configuration identifies a list of servers that compose that expected peer set of three servers. AWS fills in this value with the servers that match that `instance` tag.

4. Much like with the client configuration, we're configuring the address that we are going to bind to (`bind_addr`), the address that we are advertising (`advertise_addr`), and the address that we want to be identified as (`client_addr`). We're configuring these values using machine metadata that was set as part of the provisioning—another nice feature with Terraform and AWS.

5. Finally, we're enabling the web UI for Consul. By default, the web interface is not available, but we'll be using it occasionally throughout the book.

That completes the contents of the configuration file. The last thing we need to do is start our Consul servers!

Wahoo! Congratulate yourself for getting through all of that craziness! I hope it was as much fun for you reading it as it was for me writing it. If it makes you feel any better, *Planet Caravan* provides a great little beat for writing.

OK—by now, your Consul cluster should be up and running. Hey—*up and running*: that would be a great name for a book series! So, now that we've gotten it up, let's start taking a look around.

Accessing the Consul system

The output from the Terraform run is going to provide you with vital information for accessing the Consul cluster. This is produced as *output* from the Terraform run. Another output you'll find in your directory is an SSH key that can be used for remotely connecting to the machines, should that ever be required, as it will be here shortly. Have a look at the following code snippet:

```
Apply complete! Resources: 25 added, 0 changed, 0 destroyed.

Outputs:

CONSUL_HTTP_ADDR = "http://18.191.150.155:8500"
Consul_Server_IPs = [
  "3.129.42.175",
  "18.191.150.155",
  "3.139.81.1",
]
```

Using the preceding example, we're going to access the Consul system in two different ways: through the CLI and through the web interface.

Using the CLI

Let's start with SSH. One of the artifacts produced from that Terraform run is an SSH PEM key used to authenticate yourself with the Consul machines. The file will be called `<owner>-consul-key.pem`. Using your command-line shell, from the same place you ran Terraform, you can SSH into any of the Consul servers—for example, this one:

```
$ ssh -i rjackson-consul-key.pem ubuntu@18.191.150.155
```

Obviously, you want to replace the name of the key with your own key and the IP address of one of the Consul servers. It can be either of the three, so don't worry about picking the *right* one. You will receive a prompt with all sorts of gobbledygook (technical term) about the **Secure Hash Algorithm 256 (SHA256)** fingerprint and asking if you want to continue. Yes—you do want to continue, and you'll be on your very first Consul server!

> **Note**
>
> If you don't want to SSH into the server, you don't need to! However, in order to communicate with the Consul cluster via the CLI, you will need Consul installed locally. Don't worry—it's as simple as downloading Packer and Terraform; download a ZIP file, unzip it, and place it in your path. Consul can be downloaded from `https://www.consul.io/downloads`. If you do decide to utilize a remote CLI, just follow along, as the steps are the same.

Now that you're on the machine, please don't delete anything! Don't laugh—it happens. If you remember back when we were discussing the template file, there was a `bind_address` property that specified which address Consul was going to listen on. We are using the public address, so we want to set that as the Consul address in our environment. Don't worry—you have all of the information you need to *export* this information into your environment. I have no idea why they call it *export* when you are putting the information in, but so be it.

Within the Terraform output, you'll see a value for `CONSUL_HTTP_ADDR`. This is exactly the variable, along with that address, to set in your environment. Again using that Terraform output as an example, you can see this here:

```
$ export CONSUL_HTTP_ADDR=http://18.191.150.155:8500
```

OK—you're all set, so let's take a look at that cluster! The first thing we should do is make sure that all of our machines, servers, and clients are up and registered in the Consul system. If you don't remember how many, we have three servers and four clients. You can see all of the *members* of our cluster via the following command:

```
$ consul members
```

The output of this command provides not only the servers and clients but also the version of the agent running on each machine, along with their name, IP address, status, and data center, as illustrated in the following figure. There are some other values there too, but these are what we are most concerned with at the moment:

```
Node                   Address              Status    Type     Build   Protocol   DC    Segment
ip-192-168-100-185     18.191.150.155:8301  alive     server   1.9.0   2          dc1   <all>
ip-192-168-100-6       3.129.42.175:8301    alive     server   1.9.0   2          dc1   <all>
ip-192-168-100-74      3.139.81.1:8301      alive     server   1.9.0   2          dc1   <all>
ip-192-168-100-133     3.138.179.18:8301    alive     client   1.9.0   2          dc1   <default>
ip-192-168-100-206     3.138.135.94:8301    alive     client   1.9.0   2          dc1   <default>
ip-192-168-100-243     18.118.27.216:8301   alive     client   1.9.0   2          dc1   <default>
ip-192-168-100-30      3.15.202.61:8301     alive     client   1.9.0   2          dc1   <default>
ip-192-168-100-96      3.21.106.11:8301     alive     client   1.9.0   2          dc1   <default>
```

Figure 2.17 – consul members output

What this doesn't tell you is which server is leading the Raft. To determine that, we can execute the following command:

```
$ consul operator raft list-peers
```

This will show us which nodes are eligible to vote for the leader (at this time, all of them are, and they're still so young!). You can see the output here:

```
Node                 ID                                       Address              State      Voter   RaftProtocol
ip-192-168-100-74    2cd486d4-aeb9-b889-7ac8-2f28cb45fcf6     3.139.81.1:8300      follower   true    3
ip-192-168-100-6     50e16084-2e9d-9122-7ab6-427c94170bcf     3.129.42.175:8300    follower   true    3
ip-192-168-100-185   05abf287-cdbe-053e-2fe4-2c71af6c5aad     18.191.150.155:8300  leader     true    3
```

Figure 2.18 – Consul Raft peers

Lastly, if you want to view which services are running, you can execute the following command:

```
$ consul catalog services
```

And we'll see that our Consul cluster is already advertising a couple of services! Of course, it advertises *Consul*, but there is also that `httpd` service we set up during initialization. How cool!

Using an HTTP web interface

Accessing the Consul web interface is a *lot* easier than the command line, but even that wasn't so bad, was it? The good news is, you already have the exact **Uniform Resource Locator** (**URL**) you need to connect to. Do you remember that `CONSUL_HTTP_ADDR` value from the output? Well, that's the web interface! So, open your favorite web browser and drop `http://18.191.150.155:8500/ui` into the address bar. You'll see something very similar to this:

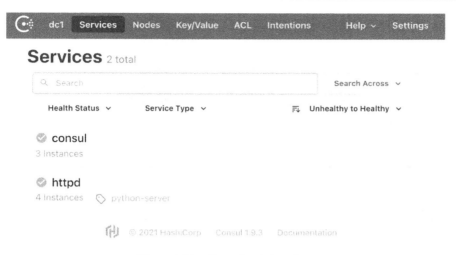

Figure 2.19 – Consul web interface

Feel free to browse around here. You can already see a list of services that are available and where those services are running. Interestingly, we can see here that **consul** is running on three instances and **httpd** is running on four. Click on the **consul** row, and you can get information about the nodes running the service, as shown in the following screenshot. These are the Consul server nodes:

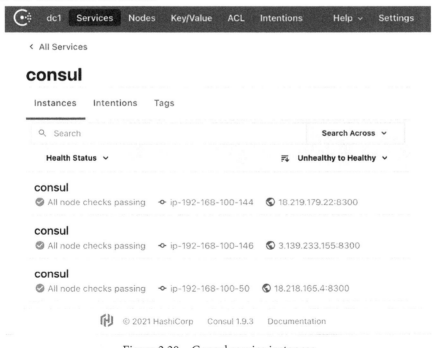

Figure 2.20 – Consul service instances

If we click on the **Nodes** tab at the top of the window, this will give us similar output to when we run our `consul members` command, as illustrated in the following screenshot:

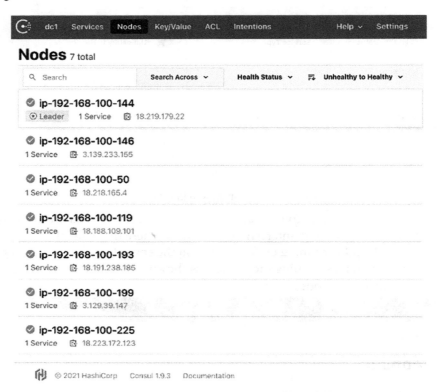

Figure 2.21 – Consul nodes listing

And there you have it: you've created an entire Consul cluster, consisting of three servers and four clients. Congratulations!

Summary

Wow—that was a long chapter, but just look at how far we've come! Utilizing Packer and Terraform, you've created your very own Consul cluster with three server nodes rafted together and four client nodes gossiping away. You have an HTTP service running and advertised on that cluster, and Consul is already monitoring its health status. And here, your high-school teachers thought you would never amount to anything…oh, wait, that was me. Well, a hearty congratulations to you.

Now that we have those communication paths throughout our system, we need to be wary of cowans and eavesdroppers. We not only have to secure all of that communication, but when we start adding more services to our cluster, we need to keep them protected as well. It's going to get thick, so before we start, grab a beverage, take a nap, or listen to some music, and prepare to secure our assets!

3

Keep It Safe, Stupid, and Secure Your Cluster!

At this point, you may be saying to yourself, *Wow, two chapters and I'm done.* Well, sorry to break it to you, but you are currently very much exposed. Yes, we've built a basic Consul cluster, consisting of three servers and four clients, but there is no security around that system. Essentially, we've built a glass house with no doors, so anybody can walk in, and anybody can see what's going on inside. Kind of creepy, isn't it?

To understand how we're going to protect our system, and more importantly our services, let's take a look at all of the areas where we are exposed:

- System communication – We have seen how all the Consul components communicate freely with each other. Anybody can join that cluster without verification, and anybody can listen in.

- Who can do what – Even if you are verified and authenticated within the cluster, we need to make sure that the scope of the actions available is tightly controlled. After all, we don't want just anybody jumping into the cluster with full access to the configuration.

- Extending Consul – With the understanding that multiple groups are going to access Consul, we want to give each of them the access they need to manage their own services, without interrupting others.

In addition to these areas, there are the basic procedures of securing the operating environment. This includes protected access to the operating system and the file structure within. These are considered basic administration procedures and aren't covered as part of this book.

With that, let's jump right in and start looking at how to start securing our glass house!

Technical requirements

With the additional complexities introduced through securing any infrastructure, we'll be working with a new repository that is specific to this chapter:

```
https://github.com/PacktPublishing/Simplifying-Service-
Management-with-Consul/tree/main/ch3
```

All the files and code are available, and we'll be utilizing Packer and Terraform again to create our working infrastructure, which means we'll also need the AWS credentials we established in *Chapter 2*, *Architecture – How Does It Work?*.

The exercises contained within this chapter have been performed within a macOS Terminal while utilizing the native SSH client. Familiarity with a command-line-based file editor is always helpful as well for viewing and editing files where necessary.

Finally, let's reiterate what was stated in *Chapter 2*, *Architecture – How Does It Work?*:

- Always perform a `terraform destroy` when you are done for the day. It does mean you'll have to rebuild as you continue the work the next day, but what a great lesson on immutable infrastructure!

- Be very careful with your AWS credentials. Never share your access key ID, and definitely don't share your secret access key. We're going to set them as environment variables so that they never show up in code, and absolutely never show up in your version control system!

Protecting the communication paths

Now, I'm not sure about you, but when I start reading documents about AES, TLS, GCM, and so on, my eyes start to gloss over. The plethora of acronyms help us chase squirrels in all directions, but that's not going to help us secure our Consul cluster. So, my goal is to explain what we need to do to secure our system communications with minimal acronyms. This reminds me of the time I had to write a paper in high school without using any of the forms of the verb *to be*. Thanks, Mr. Farley. It is important to remember that, currently, we're only going to focus on securing our clients and servers. Securing and encrypting traffic for our services will come later, so please be patient.

In the second chapter of this book, we learned about the two primary modes of communication among the Consul components: **Remote Procedure Calls** (**RPC**) and gossip. But wait – what about this Raft and consensus stuff we've read about? Well, all that communication utilizes RPC communications, and it is transmitted by utilizing TCP. Gossip is what the agents use to communicate with each other, and if you recall, it is utilizing UDP. I am calling these out because there are going to be two different methods of securing those communication paths. Let's talk about securing the gossip protocol first, simply because that's a bit easier to understand.

Securing our gossip

If you've ever seen a group of people gossiping, you know that they are usually closely gathered so that nobody else can join and listen in on their conversation. When you try to approach the group, suddenly, their conversation goes quiet, and paranoia instantly sets in. This is because you may not be a known entity to them, or they might not understand where your allegiances lie. If those people were more spread out in a larger area, due to social distancing, it would be a lot easier to listen in. This is the environment that we are working with for Consul. Especially within cloud environments, it's more difficult to keep the gossiping group close together and away from cowans and eavesdroppers.

Now that we understand that the *gossipers* can't exist as a single isolated group, what is the best way to ensure that what they are talking about can't be heard, or at least understood, by others? Let's put a twist into our story (not a piece of citrus for our drink). Let's assume you find yourself in the midst of groups of people speaking a language you aren't familiar with. I can imagine most of us have been in that situation – even if they are speaking the same core language, you might not understand some of the words they're speaking. My wife, having grown up in Indiana, experienced this when we moved to the Boston area, but that's another story. Within that environment, you may hear that people are speaking, but you're unable to understand what they are saying. This is a great way to keep communication open, but safe from those not entitled to the message. This is exactly what we are doing to secure our gossiping agents.

Consul provides a method of generating a special language, defined by an *encryption key*. This message is the same as what we created, but the language that's used is different. This key is distributed to all the agents within the Consul cluster. Even if you have multiple clusters communicating with each other, they all need to use the same language, and therefore the same key.

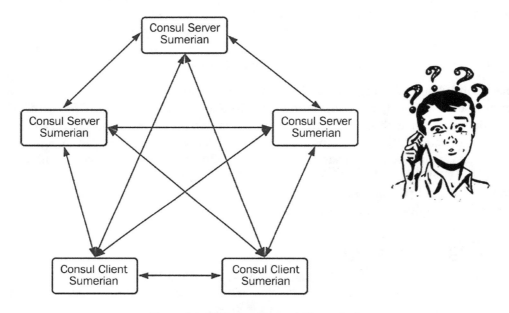

Figure 3.1 – Who understands Sumerian?

At this point, you might be thinking *but Rob, eventually people can learn a new language and decipher the message*. Right you are, and although it may take some of us more time to learn a different language than others, eventually, if we work hard enough, we can figure it out. This is the same for every encryption key, password, and certificate, and is the main reason why it's important to periodically change and update these secure items. Essentially, as those around the gossipers learn the new language, we need to change the language! When we start talking about tens, hundreds, or thousands of gossipers in our group, it can be very challenging to have everybody learn the new language, without missing messages. Thankfully, Consul can manage a great deal of this for you, and we'll see how that's done when we deploy our new secure system.

Securing our calls

Unlike gossip, when we call people (unless, of course, it's a conference call), the communications aren't broadcast to a distributed group of people. That conversation is usually person to person, with some level of verification of who you are speaking with. This can be the phone number, the source number, a familiar voice, and so on. When we look at the communication among our Consul agents, the process is very similar. However, we never know when somebody else might be listening in on that conversation. Therefore, there are three steps to securing our calls: validating who is calling, validating who is being called, and encrypting the messages between them.

Figure 3.2 – Secure versus insecure communications

The way we do this with Consul isn't that different from what you do nearly every day when you open a web browser, log in to your bank account, and send lots of money to me. When you do this, your web browser validates that it is connected to the proper website for your bank. This validation comes from a certificate that is signed by some official authoritative entity, ensuring your browser that the site is valid. If you've ever seen a pop-up window from your browser that says the certificate can't be authenticated, that means it wasn't signed by a known and accepted authority. As that certificate is exchanged, the browser then knows how to encrypt traffic so that the banking website can decipher it, and similarly, your browser knows how to decipher the messages coming back from the bank. However, there is one slight variation to what we do with Consul.

Let's take the scenario again where you are accessing your bank's website from your browser. The bank's website doesn't really care where you are (in most instances, at least). With Consul, the agent making the call not only validates that it's calling a valid entity, but the one being called also ensures that the caller is valid. Let's dig into this a bit deeper, but don't worry – we won't be hitting the water with this dig.

We're going to be talking a lot about keys and certificates within this section. If you've never worked in any detail with **Transport Layer Security** (TLS), it can get a bit confusing, so I'm going to be using some notation to identify servers, clients, and private and public keys, based on the characters of the long-running animated series *The Simpsons*. For example, we're going to have three servers: Bart, Lisa, and Maggie. For these three servers, there needs to be some sort of trusted authority to identify Bart, Lisa, and Maggie; we'll call our authority Marge.

All three of our servers require some sort of method to transmit and receive messages that only they can understand. If you remember our telephone analogy, they all need to validate who is calling, who is being called, and encrypt the messages between them. All of this is done using certificates that are made especially for them. Now, if Bart created his own certificates and distributed those to Lisa and Maggie, they could share secret messages. However, what would prevent Homer, who really shouldn't be privy to the messages, from jumping into the group, sharing his own certificates, and seeing these secret messages? For this, we need Marge, some sort of centralized authority to ensure that everybody is using the certificates that she authorizes. We can call her a Certificate Authority (CA).

Using her awesome power, Marge produces two certificates for Bart, two for Lisa, and two for Maggie. These certificates have special names: a *public key* and a *private key*. The public key confirms their identity and enables them to encrypt messages to the others, while the private key enables them to decrypt the messages they receive. Optionally, Marge can create specific constraints, such as how long the certificates will last, and which family names are allowed to utilize the certificate. For example, if somebody walks in and their last name isn't *Simpson*, then they won't be able to use Marge's certificates.

Marge produces these certificates using two certificates of her own; one that identifies her authority, which we'll call *porkchop*, and another secret ingredient we'll call *salt*. Nobody else knows about Marge's private secret, for if they did, they could use salt in their own certificates, thereby compromising Marge's authority. To summarize, let's look at the certificates that we have:

- `porkchop`: A certificate that identifies Marge as authoritative
- `salt`: Marge's secret ingredient that she uses to produce certificates
- `bart.public_key`: A certificate that Bart uses to identify himself and encrypt secret messages to Lisa and Maggie
- `bart.private_key`: A certificate that only Bart has, and that he can use to decrypt messages from Lisa and Maggie

- `lisa.public_key`: A certificate that Lisa uses to identify herself and encrypt secret messages to Bart and Maggie

- `lisa.private_key`: A certificate that only Lisa has, and that she can use to decrypt messages from Bart and Maggie

- `maggie.public_key`: A certificate that Maggie uses to identify herself and encrypt secret messages to Bart and Lisa

- `maggie.private_key`: A certificate that only Maggie has, and that she can use to decrypt messages from Bart and Lisa

If Lisa wants to send a message to Maggie, she encrypts the message using her own special public key. Lisa sends that message to Maggie, who can open the message using her own private key, and validates that message using the porkchop that Marge provided. Now, let's say Homer can create his own public key and encrypt his own message to Maggie. Well, that key didn't use Marge's authority, nor does it have her secret ingredient. Therefore, Maggie knows to reject the message, and sad Homer heads to Moe's.

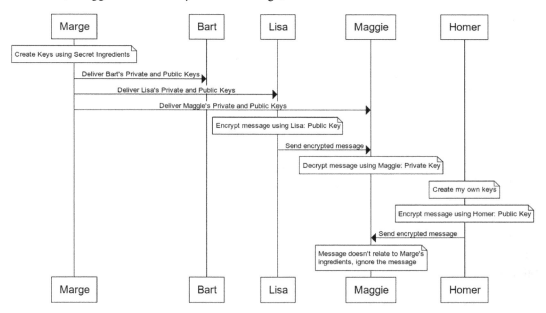

Figure 3.3 – Certificate and key distribution

This is exactly how we secure the calls between our Consul machines. Every node in the cluster has its own public key, private key, and an authoritative key provided by some centralized CA. For our exercises, we'll be using Consul as our CA. This is just to help us be as independent as possible, but this shouldn't be done in a production environment. In the real world, your company would have a CA and somebody in the security team would provide these keys to you (as well as new keys as the certificates are rotated). If your company has embraced automation, it's possible to generate these keys by utilizing HashiCorp's Vault if, of course, you are authorized to do so.

Now that we've seen how to secure both the gossip among our nodes and their direct messages, let's look at what else we can do to secure our system.

Controlling access with access control

At the heart of Consul's functionality is a distributed key/value store. Through the gossip protocol, anything that one node learns can be shared with other nodes in the cluster. This not only applies to our three main use cases, but often, operators will utilize Consul to distribute configuration parameters that the associated components can utilize. This can be very helpful if you want to quickly change the logging level on all machines. This is especially the reason why, in any production deployment, the importance of securing Consul can't be understated. Utilizing the TLS keys, as we discussed in the previous section, we can validate and secure the communications between the clients and the servers. If we didn't do this, it would be possible for somebody to pretend to be part of the cluster.

The access control system within Consul provides an additional level of control regarding what each node can do on the network – hence the term *access control*. If you've ever had to deal with **Access Control Lists (ACLs)** on routing equipment or network firewall rules, you may already be familiar with the concept of access control. However, with Consul, the rules can be very fine-grained and detailed. Of course, this means they can grow to be very complex, increasing the need for some level of infrastructure as code (such as Terraform) or another automation tool. Before we get into that, though, let's understand a bit more about what the Consul access control system consists of.

When we discuss access control in the realm of Consul, there are three main pieces that we need to understand: rules, policies, and tokens. Optionally, we can create roles for groups of policies as well, targeting groups of nodes, services, or data centers that the policy should be applied to. We've talked about ensuring identification a great deal in this chapter. After all, this chapter is about security, and a big part of security is ensuring identity. This is why secret handshakes and passes were invented. For Consul access control, that identity is provided in the form of a token that is generated by some authoritative entity. We're going to be using Consul for that authority; however, centrally managing these tokens in a system such as HashiCorp's Vault would be preferred, not just because I'm currently employed by HashiCorp, but primarily because utilizing Vault provides the operator with a centralized and secure method of maintaining and rotating these tokens. OK, enough about tokens… what can a token do for me?

When somebody uses a token to access Consul, the token is linked with a policy, which contains one or more rules. As the policy is simply a group of rules, we're going to focus on the rules that we can apply.

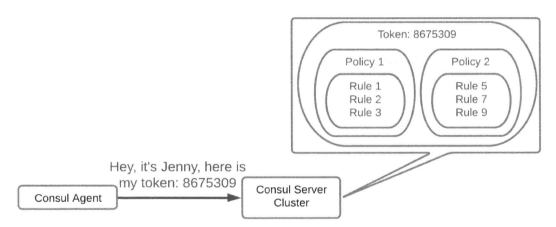

Figure 3.4 – Tokens, policies, and rules

Nearly every rule contains three parts: a rule type, a variable that defines where that rule is applied, and a control level for a rule. There are some special cases where the variable is not necessary, but we'll address those later. There is also always a default policy, which is either *deny* or *allow* based on your security posture. Now that I've mentioned *posture*, I sat up straight in my chair… did you? Extreme care must be taken when defining that default policy. If the default is *deny*, then every operation, service, node, and so on needs explicit permission to perform some function. If the default policy is *allow*, then you need to specify what you are going to block. Certainly, the latter appears much simpler to implement. However, from a security perspective, it is a great deal easier to only identify the few things you want to permit, rather than try to identify and prevent all those things you need to block (which changes constantly). So, moving forward, we're going to assume our default policy is to deny. You may hate me for this now, but you'll appreciate it later.

You can guess the control levels for Consul rules, but just in case you can't, they are as follows:

- **read**: Read access to a particular resource grants those using the token the ability to read the values associated with the resource.

- **write**: Write access to a particular resource grants those using the token the ability to update or create values associated with the resource. If your token grants write access to the resource, you also have read access.

- **deny**: Smackdown – you have no permission to read or write values associated with the identified resource. As we'll be using a default deny policy, we won't need this level.

- **list**: This is a special level that grants those using the token the ability to list all associated parameters. The only place this is applicable is within the Consul key/value store, where the bearer of the associated token can list all key/value combinations at a particular path prefix.

As I brutally learned in math class, whenever you are looking at areas like this, the order of precedence is critical. If there are any overlapping rules, deny has priority overall, and the ability to write has priority over read-only. This should be a given, but the fact that deny has priority over everything should be well understood. Furthermore, if you have *write* access within a particular rule, you automatically get list access. If you have *list* access, you automatically have *read* access, so the level of control is somewhat nested. I find one way to easily remember this is through the following diagram:

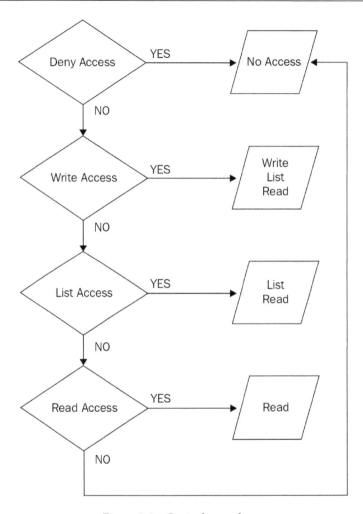

Figure 3.5 – Control precedence

Now that we know *what* we can do with individual rules, let's look at *where* we can (or can't) perform these actions… which brings us to rule types. The following is a list of all the rule types currently supported. Although we will only be using a few of them during this exercise, I wanted to list them all simply for completeness:

- **ACL resource**: Controls the management of ACL rules.

- **Agent**: This rule type controls aspects related to managing the Consul agent, such as joining a cluster, listing members, and agent configuration. Agents really should have write access to their own nodes; otherwise, they won't be able to register their own service information with the catalog.

- **Event**: This is specifically related to the ability to read and write Consul events. We won't be covering events in this book, but they can be utilized to exercise custom actions based on changes within the Consul system, including service availability.

- **Key/Value**: Control access to the distributed key/value store within Consul. Rules can apply to the ability to read, write, or list sets of keys and their associated values.

- **Sentinel**: No relation to the incredible song by Judas Priest, the Sentinel rule allows us to write just about any rule possible based on the Consul structure. Sentinel is an Enterprise feature, and quite advanced, so we won't be covering it in this book.

- **Keyring**: This is one of the rule types that requires no variable to be applied, as it is specific to the keyring function. As the keyring is what controls the encryption key for the gossip protocol, access to this area should be tightly controlled.

- **Node**: This is an important one. It controls node-level data such as the services available in the catalog and node health. This also operates as a filter for agent-level operations to the subset of cluster members permitted.

- **Operator**: This is another rule type that requires no variable to be applied and addresses the operator-level operations within Consul, such as auto-pilot or Raft specifics. Like the keyring, access to this area should be tightly controlled.

- **Prepared query**: Consul offers the ability to create templates that are useful for more complex service queries within the system, including node location, metadata, tags, and so on. We will not be covering prepared queries as part of this book.

- **Service**: Here is another important one to understand. This rule type governs service-level information, including access to the service catalog and service health status.

- **Session**: Consul enables you to allocate and release a lock on values within the key/value store. This can be done in performed sessions and will not be addressed as part of this book.

- **Namespace**: Namespaces are another Enterprise-level feature and enable the subdivision of Consul clusters. This rule type allows you to create more comprehensive rules consisting of multiple rule types for an entire namespace. Although we won't be covering namespaces much in this book, it is very important to understand that several rule types, including the operator and keyring rules, can't be applied to namespaces. These functions are global in scale and only operate on the default namespace.

As you can see, there are a lot of areas where we can apply ACLs to manage what humans, machines, or cyborgs can do to access our services. This also provides a plethora of areas to control the individual nodes.

When we are talking about security, we often use the term *least privilege*. Plainly speaking, or writing, this means that instead of giving a child a pile of candy and telling them what they can't eat, we start with the child having no candy and providing only what they can eat. This is why, earlier in this section, we talked about instituting a default deny policy within Consul. If we utilize this as our standard security rule, especially with the certificates we discussed in the previous section, we can further tighten our control of the individual nodes. This drastically reduces the possibility of any cowans or eavesdroppers gaining access to, or listening in on, our cluster operations – for if you don't have the secret handshake, you will not pass.

Now that we've addressed all of that, let's take a look at what a rule looks like. We're going to look at two rules here, both of which we'll actually use when we start getting our hands on the system.

This first rule applies a minimal policy for a Consul client to perform its operations on itself and alert the Consul server of any changes to the infrastructure:

```
node "ip-192-168-100-11" {
  policy = "write"
}
service "consul" {
  policy = "write"
}
agent_prefix "ip" {
  policy = "write"
}
```

Within this policy, we have three main rules that appear to be quite simple, but within a least privileged model, their role in society is critically important.

The first rule within the policy utilizes the node rule type. Without this rule, the node can't update the server cluster on what services it sees and, more importantly, the health and location of the node. If a client attempts to join the cluster, without the token associated with this rule, the server cluster won't even recognize it as a client. Note that because the node rule only applies to ip-192-168-100-11, this rule only applies to the actions for this particular node, as that is its hostname. Any other node using the token associated with this policy would not be recognized by the Consul server cluster.

The next rule, which is of the `service` type, allows us to update the service catalog on the node itself. This is particular to the services on the node itself, not the overall service catalog hosted by the Consul server cluster. To update the centralized service catalog, we'll need the agent rule, which we'll talk about next. Note that with this rule, we can `write` to the service named `consul` to manage service changes. If we only wanted to read the status or availability of the service, we could use a `read` directive.

The last rule utilizes the `agent` rule type. This is similar in importance to the node rule type, where it controls very fundamental aspects of the functionality of the node (although in this case, it's pertinent to the agent). This rule allows the bearer of the associated token to perform agent-level functions such as updating the service catalog... which is kind of important since Consul is all about services. One difference that you might have picked up with this particular rule definition is that instead of pointing to a particular agent, it utilizes the slight variation called `agent_prefix`. As you might have guessed, this applies to any agent with a hostname that starts with the prefix `ip`. This provides some flexibility, allowing you to apply rules to groups of agents, as opposed to writing a rule for each individual agent. This does, however, expose us a bit because the rule isn't locked down to one particular agent. One thing that can be done to help here is to have two agent rules – one specific to the host that it resides on, enabling it to write to agent operations, and another to read agent operations from other nodes:

```
node "ip-192-168-100-11" {
  policy = "write"
}
agent "ip-192-168-100-11" {
  policy = "write"
}
agent_prefix "ip" {
  policy = "read"
}
```

This was a very constricted set of policy rules applicable to that particular agent. The following policy opens things up a bit and is more applicable to the server cluster, affording it the ability to do things such as write ACLs, events, and more:

```
acl = "write"
agent_prefix "" {
    policy = "write"
}
event_prefix "" {
```

```
        policy = "write"
}
key_prefix "" {
        policy = "write"
}
keyring = "write"
node_prefix "" {
        policy = "write"
}
operator = "write"
query_prefix "" {
        policy = "write"
}
service_prefix "" {
        policy = "write"
        intentions = "write"
}
session_prefix "" {
        policy = "write"
}
```

When we *bootstrap* a cluster to apply ACLs, this is a similar policy to what would be created and is analogous to *admin*-level privileges. Note here that not all the rule types apply to any particular node. This is useful for getting started, but in a production environment, you would definitely want to lock things down a bit tighter.

One thing to note here is that we are using a lot of prefix type rules within this list. With these rules, you might be asking yourself, what if I have multiple prefixes with different capacities for each one? This is a great question, and it depends on the specificity of the rule. The most specific area of applying the rule, or segment, takes precedence. For example, let's take a look at the following scenario:

```
agent_prefix "ip-192" {
        policy = "write"
}
agent_prefix "ip" {
        policy = "read"
}
```

Here, the bearer of the associated token to this policy would be able to write to any hostname starting with `ip-192`, but would only have read-only access to any hostname starting with `ip`.

As I hope you can see, the flexibility of the ACLs within Consul allows us to define and enforce incredibly specific rules on our system. We can apply a variety of controls to exact areas of the Consul functionality. Some of these rules are crucial for any functional system (such as node or agent types), but even those should be controlled so that they're only applied to specific segments (hostnames, domains, and so on). Next, we're going to look at what we can do to manage the humans that have access to our Consul cluster and track what they are doing, just like Big Brother.

As we have already seen, the ability to use Terraform to create and manage infrastructure simplifies a lot of this complexity. When you start working with Consul clusters containing hundreds, thousands, and tens of thousands of nodes (and yes, clusters of that size do exist), the convenience of using infrastructure as code becomes more of a necessity. If you're tired of reading and want to see how all these security features come together, just hang tight – we'll get there soon. First, we need to take a quick look at audit logging within the Consul system, a critical component for any enterprise.

Role up your identity

As part of enabling production for a service, we need to work with who is going to access the system, and what they can do. This is typically referred to as **Role-Based Access Control**, or **RBAC**. We also want to be able to maintain an audit trail for the system. This last part is becoming increasingly important as we get out of that trusted walled garden network and into the cloud. I can't tell you how many times, even in a controlled environment, I had to go back and figure out what changed and when that broke my system! In the interests of full disclosure, an Enterprise-level license for Consul is required to implement this level of functionality.

The actual process of logging in to the Consul system can certainly be managed by using ACLs and then treating any human like a machine by giving them their own particular access token. However, these tokens aren't as simple as Jean Valjean's 24601. So, inevitably, somebody will end up writing their token down on a sticky note attached to their laptop, or better yet, emailing it to their friends and family with the latest cat meme. Furthermore, this only works until you have some engineer (could it be you?) standing up in front of the company exclaiming *I am not a number, I am a free man!* Cue the maniacal laugh emanating from the offices of upper management. A much better solution is to tie Consul into some centralized method of authentication and authorization. For this, Consul Enterprise utilizes one of the growing standards for centralized authentication and authorization – **OpenID Connect (OIDC)**.

Utilizing OIDC, Consul can validate users attempting to access Consul by querying one of the many OIDC providers (most enterprise environments utilize at least one). If you've ever logged in to a service and were redirected to another authentication page or popup for your Google, Yahoo, Microsoft, PayPal, or Amazon credentials, it is likely that you've utilized OIDC for that authentication. Other popular OIDC providers for enterprise environments include OKTA and Auth0. This process not only validates and authenticates the user, but it can also map that user to a set of authorization policies to control access to the system. After all, if you ever have to give a manager access to your Consul cluster, you certainly don't want them having administrative privileges! So, let's walk through this with another lovely example.

As part of this story, we need to assume that our greater employer has implemented OIDC and that each of us has our own credentials to authenticate with that OIDC provider. Along with our credentials, we all have specific metadata that is associated with our position in the company:

Credentials		Metadata		
Username	**Password**	**First Name**	**Last Name**	**Position**
sharris	************	Steve	Harris	Bassist
bdickinson	************	Bruce	Dickinson	Singer
dmurray	************	Dave	Murray	Guitarist
asmith	************	Adrian	Smith	Guitarist
nmcbrain	************	Nicko	McBrain	Drummer

Figure 3.6 – OIDC accounts and metadata

In this example, everybody has their own unique position, except for Adrian and Dave. They are both in the *guitarist* group, which we're going to map to a role within Consul. Recall that a *role* consists of a group of ACL policies within Consul. When we configure Consul to utilize this information, we create what's called a **claim mapping**. This maps the metadata that's been provisioned within the OIDC provider to variables within Consul, which can then be used to associate that metadata with a particular *role*. For example, the following configuration associates specific mappings for the first and last name, and the position metadata is associated with a list in Consul called band_positions:

```
{
    "OIDCDiscoveryURL": "https://<OIDC_URL>/",
    "OIDCClientID": "<OIDC_CLIENT_ID>",
```

```
"OIDCClientSecret": "<OIDC_CLIENT_SECRET>",
"BoundAudiences": ["<OIDC_CLIENT_ID>"],
"AllowedRedirectURIs": [
  "http://localhost:8550/oidc/callback",
  "http://localhost:8500/ui/oidc/callback"
],
"ClaimMappings": {
  "user.metadata.first_name": "first_name",
  "user.metadata.last_name": "last_name"
},
"ListClaimMappings": {
  "user.metadata.position ": "band_positions"
}
}
```

This configuration simply allows Consul to communicate with the OIDC provider and pull specific account information (first name, last name, and position). However, this information means absolutely nothing unless we configure Consul to associate that metadata with a specific role. Let's pretend we have a role within Consul for guitarists that consists of two policies. We can create that role using the following example command:

```
$ consul acl role create -name guitarists -policy-name rhythm
-policy-name lead
```

Wonderful! We have created our accounts in our OIDC provider, and we have a role defined within Consul that associates specific attributes of the account with a set of policies. However, how do we link these two pieces of information? This can be done through what is called a **binding rule**. Now, we haven't talked about these before, but a binding rule associates a method of authentication with a particular role. Binding rules can be applied to application authentication (for example, if you wanted to validate authentication using a Java Web Token), but our focus here is to authenticate humans. Specifically, we want humans that have a band position of guitarist to have access to the guitarists Consul role. Note a slight difference in spelling (guitarist versus guitarists). This was done to differentiate between terms through the process.

Enough talk about our binding rules. If we wanted to create one, we could use the following example:

```
$ consul acl binding-rule create -method=<oidc_name> -bind-
type=role -bind-name=guitarists -selector='guitarist in list.
band_positions'
```

Here, we are using the name of the OIDC provider that was given when we configured the OIDC provider within Consul. We are stating that anybody who logs in to Consul using that method needs to be authenticated using the associated OIDC provider. When we receive the metadata back for the user attempting to authenticate, we are looking in the list of `band_positions`, and if we find a guitarist, we can assign the role (and therefore all the associated policies) of `guitarists` to those users. Now, why would we go through all this work when all I need to do is create tokens and give those tokens to users?

Let's assume we've gone through the process of creating a role for guitarists, and that the role has its own token. We can give that same token to Adrian and Dave, and they can authenticate with Consul (using the CLI or web interface) using that token. Here is problem number one; how can we differentiate between Adrian and Dave when we are tracking audit logs? It might not be too bad when we only have one user, but when Jannick (another guitarist) is added to the band, it becomes even more difficult to differentiate users (although their guitar styles are vastly different).

Now, I admit that all of this can be confusing, so let me lay out the sequence:

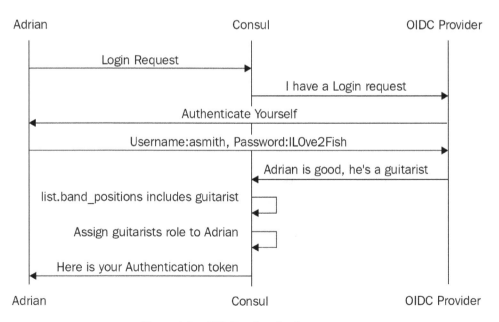

Figure 3.7 – OIDC authentication process

I had mentioned the necessity to differentiate between Adrian, Dave, and now Jannick, as they log in to the Consul system. This is to differentiate between actions performed within the audit log by each individual. This isn't necessarily to stand back and point fingers, but more to ensure that the process is followed and changes are tracked throughout the system.

Enabling the audit log is pretty simple within Consul servers. When enabled, Consul will spit out entries to track who did what and when, and it all gets written to a file. Don't worry – secret information isn't included, which gives greater importance to enabling authentication methods such as OIDC to uniquely distinguish users. The logs are written to a file in JSON format, which can be ingested into most log analysis utilities such as DataDog, Splunk, and Elasticsearch. When configuring the logging, however, be sure to set limits on the number of log files and the rotation frequency… nobody ever likes to admit that they ran out of space due to logging!

```
audit {
  enabled = true
  sink "consul_audit" {
    type   = "file"
    format = "json"
    path   = "data/audit/audit.json"
    delivery_guarantee = "best-effort"
    rotate_duration = "24h"
    rotate_max_files = 15
    rotate_bytes = 25165824
  }
}
```

The output of the log file identifies several important pieces of information, specifically the timestamp and the token accessor. This accessor can be linked back to the individual tokens and roles (and therefore users) that accessed the system. I personally find the source IP address to be very interesting in audit logs, especially when I can do geographic association and determine who is accessing the system from various parts of the world. Of course, the entry includes what they did with the system as well!

```
{
  "created_at": "2020-08-25T10:35:23.258765383Z",
  "event_type": "audit",
  "payload": {
    "id": "3a880e2b-3f43-5d4d-97a6-26aec5eacc0c",
```

```
        "version": "1",
        "type": "HTTPEvent",
        "timestamp": "2020-08-25T10:35:23.258548744Z",
        "auth": {
            "accessor_id": "48162132-6b28-de21-aee9-6903d2024f2e",
            "description": "token created via OIDC login",
            "create_time": "2020-08-25T10:35:23.158336091Z"
        },
        "request": {
            "operation": "GET",
            "endpoint": "/v1/catalog/datacenters",
            "remote_addr": "172.17.0.1:48516",
            "user_agent": "Mozilla/5.0 (Macintosh; Intel Mac
OS X 10_14_6) AppleWebKit/537.36 (KHTML, like Gecko)
Chrome/84.0.4147.125 Safari/537.36",
            "host": "192.168.64.243:31394"
        },
        "stage": "OperationStart"
    }
}
```

As you can see, enabling these audit logs is a crucial aspect of operating and managing the Consul system within any enterprise. Because we are simply practitioners learning the system, it isn't as interesting (we should know what we did and when, although I admit I probably use the Bash history more than I should). However, as Consul is rolled out to enterprise environments, tracking these operations is as critical as ensuring the system is actually running! Speaking of running, I hope you're ready. I would strongly advise at this point that you get up, take a walk, or stretch your legs, because it's about to get pretty thick.

Securing your Consul cluster

Before we start securing your cluster, I need to set forth a word of caution. Things can get a little tricky when you work with security enforcement, especially when you accidentally lock yourself out of your own cluster. Luckily, we're just working with a simple development system here. However, in a production environment, extreme care must be taken. I've tried to automate as much as possible, but at any point in this process, you can destroy the entire infrastructure set and rebuild it. Ah, the wonder and beauty of Terraform and ephemeral infrastructure!

All the code for this section can be found in the following GitHub repository:

```
https://github.com/PacktPublishing/Simplifying-Service-
Management-with-Consul/tree/main/ch3
```

Creating the image

If you've already forked or downloaded the repository, you should have the corresponding folder for this chapter. This is where we'll be doing all our work. We are going to be building a new image for this chapter that includes all the certificates that have been generated for you.

As we build the system and look at the code, we're going to focus on the differences between what we did in *Chapter 2, Architecture – How Does It Work?*, and what we are doing here, simply for brevity. Unless you actually shut down your machine or closed your Terminal, you should still have your AWS keys: AWS_ACCESS_KEY_ID and AWS_SECRET_ACCESS_KEY. To check whether you still have these set properly, run the following command:

```
$ env | grep AWS
AWS_ACCESS_KEY_ID=AKIA26CFBA32D6JFVJJB
AWS_SECRET_ACCESS_KEY=NYaSFhAzZqOq4bqSGJda/Y22Guc3p...
```

Since you are all Packer experts by now, we're going to kick off the build so that we can review what we are doing while the image is created. To do that, you need to be in the proper folder:

```
$ cd $github_root/Consul-Up-and-Running/ch3/Image-Creation
```

Now, kick off your Packer building using the following command:

```
packer build -var-file="variables.pkrvars.hcl" .
```

Now, while the Christmas lights are spewing across your screen, let's take a look at the differences between what you are building now and what you built in *Chapter 2, Architecture – How Does It Work?*.

The first thing you might recognize is that in the source definition, we have slightly changed the name of the image to be used. This is just because we only want to be using this image to understand the security aspects of Consul. As we move forward, we will be using the insecure image. *But Rob, why would it be insecure?* I've been asking this myself nearly all my life! As we will see, once we secure our cluster, it becomes more much difficult to utilize and navigate. This is by design because, in a production environment, you don't want to make it easier for those nefarious hackers:

```
source "amazon-ebs" "ubuntu-image" {
  ami_name = "${var.owner}-secure-consul-{{timestamp}}"
  region = "${var.aws_region}"
  instance_type = var.aws_instance_type
  tags = {
    Name = "${var.owner}-secure-consul"
  }
}
```

We aren't really changing things that drastically, only slightly modifying the name to identify it as the secure image base. One of the small changes for the secure image is the presence of the CA file, `consul-agent-ca.pem`. Recall that the CA confirms that the certificate being used is valid. Marge needs to be sure that her kids are all using valid certificates! We could use a provisioner to move this file to the machines when we run the Terraform script; however, every machine needs the CA file, so we might as well pack that into the image:

```
provisioner "file" {
  source      = "../files/consul-agent-ca.pem"
  destination = "/tmp/consul-agent-ca.pem"
}
```

We're also going to be moving that file to the proper location within Consul as part of our shell provisioner:

```
provisioner "shell"{
  inline = [
    "sudo /usr/local/bin/consul -autocomplete-install",
    "sudo useradd --system --home /etc/consul/consul.d --shell /bin/false consul",
    "sudo mkdir /etc/consul /etc/consul/consul.d /etc/consul/logs /var/lib/consul/ /var/run/consul/",
    "sudo chown -R consul:consul /etc/consul /var/lib/consul/
```

```
/var/run/consul/",
      "sudo chmod -R a+r /etc/consul/logs/",
      "sudo mv /tmp/consul.service /etc/systemd/system/consul.
service",
      "sudo mv /tmp/consul-common.hcl /etc/consul/consul.d/
consul-common.hcl",
      "sudo mv /tmp/consul-agent-ca.pem /etc/consul/consul.d/
consul-agent-ca.pem"
   ]
 }
```

And that's it! These are seemingly small changes, but we should get used to packing our commonalities into the image files to manage how we build out our systems in that crazy place called the *real world*.

Next, we're going to build our Consul system. Again, this process is very similar to what we did previously. We're going to kick off `terraform apply`, and while that is running, we'll take a look at the differences.

Building the Consul cluster

Now that we've built the image, we can use Terraform to build the cluster itself. All the Terraform code exists in this chapter's root folder, so let's go there:

```
$ cd $github_root/Consul-Up-and-Running/ch3
```

As you may recall, the `terraform.tfvars` file doesn't get copied over from what you configured in *Chapter 2, Architecture – How Does It Work?*. So, you'll have to make sure you edit it so that it matches your Chapter 2 file (or just copy the Chapter 2 `terraform.tfvars` file to the Chapter 3 directory):

Terraform.tfvars

```
owner         = "rjackson"
aws_region    = "us-east-2"
instance_type = "t2.small"
```

Now that we've confirmed our variables are to our liking, let's start firing up our infrastructure. Remember that to get started, Terraform needs to download any provider-specific data necessary. This can be done via the following command:

```
$ terraform init
```

Next, we're going to run our plan so that we can see what changes Terraform is going to make. This will be all new infrastructure, so you should just see a bunch of green plus signs:

```
$ terraform plan
```

Let's start building our infrastructure now! Once again, by living dangerously (not recommended in production), we can auto-approve our `apply` command:

```
$ terraform apply --auto-approve
```

OK; let's now look at what we are building. Within the definition of the security groups within the `main.tf` file, we did make a slight modification to include access to port `8501`. This is the port that Consul is using for secure communication:

```
ingress {
    from_port   = 8501
    to_port     = 8501
    protocol    = "tcp"
    cidr_blocks = ["0.0.0.0/0"]
}
```

The majority of the changes, however, are where we are creating our AWS server instances. The big difference here is that we are pushing our private and public keys for each server out to the system. Please note that I'm using very generic certificates that were created by Consul. In most instances, you'll want to use specific certificates for your environment. As the process of creating and distributing the certificates may cause even the most stable person to lose their mind, I've done a lot of that for you. If you wish to follow the process of creating the certificates, the following command must be run for every server. These commands are simply for informative purposes as certificates have already been embedded in the infrastructure:

```
consul tls cert create
```

You can use this command to create server and client certificates, and even a certificate for the CLI to access the Consul system. They would all use the same CA file that is also produced by the command. The output would be as follows:

```
consul tls cert create -server
==> WARNING: Server Certificates grants authority to become a
    server and access all state in the cluster including root
keys
```

```
    and all ACL tokens. Do not distribute them to production
hosts
    that are not server nodes. Store them as securely as CA
keys.
==> Using consul-ca.pem and consul-ca-key.pem
==> Saved dc1-server-consul-0.pem
==> Saved dc1-server-consul-0-key.pem
```

This command would need to be executed for each of the three servers in the server cluster, and the files would need to be transferred over. For simplicity's sake, I've created the certificates ahead of time and placed them in their proper positions. A big part of the changes that have been made to the Terraform script in the `main.tf` file is due to the placement of these files. It can't be stressed enough that in a production environment, you should be using certificates provided by your security team (or through automation with HashiCorp's Vault):

```
main.tfresource aws_instance "consul-server" {
  count                       = 3
  ami                         = data.aws_ami.an_image.id
  instance_type               = var.instance_type
  key_name                    = aws_key_pair.consul_key.key_
name
  associate_public_ip_address = true
  subnet_id                   = aws_subnet.consul-demo.id
  vpc_security_group_ids      = [aws_security_group.consul-
demo.id]
  iam_instance_profile        = aws_iam_instance_profile.
instance_profile.name
  user_data = templatefile("files/server_template.tpl", {
server_name_tag = "${var.owner}-consul-server-instance",
server_number = count.index})
  tags = {
    Name    = "${var.owner}-consul-server-instance"
    Owner = var.owner
    Instance = "${var.owner}-consul-server-instance-${count.
index}"
  }
  provisioner "file" {
    source      = "files/dc1-server-consul-${count.index}.pem"
    destination = "/tmp/dc1-server-consul-${count.index}.pem"
```

```
  }
  provisioner "file" {
    source       = "files/dc1-server-consul-${count.index}-key.
pem"
    destination = "/tmp/dc1-server-consul-${count.index}-key.
pem"
  }
  provisioner "file" {
    source       = "files/server_acl.hcl"
    destination = "/tmp/server_acl.hcl"
  }
  provisioner "remote-exec" {
    inline = [
      "sudo mv /tmp/*.pem /etc/consul/consul.d/"
    ]
  }
  connection {
    type        = "ssh"
    user        = "ubuntu"
    private_key = local_file.private_key.content
    host        = self.public_ip
  }
}
```

The main point to focus on is the slight change in the user_data script. Instead of simply passing the name of the server through the configuration, we've also included the index. This is because the files that we are using for the certificates all follow the same naming structure, with the index embedded in the filename. These are all visible within the files folder ($github_root/ch3/files):

```
files/consul-agent-ca-key.pem
files/consul-agent-ca.pem
files/dc1-server-consul-0-key.pem
files/dc1-server-consul-0.pem
files/dc1-server-consul-1-key.pem
files/dc1-server-consul-1.pem
files/dc1-server-consul-2-key.pem
files/dc1-server-consul-2.pem
```

The `consul-agent-ca-key.pem` file is the private key for the CA (Consul is our CA). The public CA file, `consul-agent-ca.pem`, was pushed to all the machines as part of the image build process. By using the provisioners within the `aws_instance` resource, we are transferring the remaining to each of the three servers (0, 1, and 2) and placing them in the `/etc/consul/consul.d/` directory.

Utilizing the file provisioner, we are also adding an ACL configuration file to the server. This is just to help us when the time comes for us to start playing with ACLs. When we provision our clients, there are a couple of small changes we have to make. First, we must move a new configuration file to the clients to help with the ACL configuration. We're also adding the CA file to the configuration:

```
resource null_resource "provisioning-clients" {
  for_each = { for client in aws_instance.consul-client :
client.tags.Name => client }
  provisioner "file" {
    source      = "files/httpd.json"
    destination = "/tmp/httpd.json"
  }
  provisioner "file" {
    source      = "files/client_acl.hcl"
    destination = "/tmp/client_acl.hcl"
  }
  provisioner "remote-exec" {
    inline = [
      "sudo cat << EOF > /tmp/consul-client.hcl",
      "advertise_addr = \"${each.value.public_ip}\"",
      "server = false",
      "enable_script_checks = true",
      "bind_addr = \"${each.value.private_ip}\"",
      "retry_join = [\"${aws_instance.consul-server[0].public_
ip}\",\"${aws_instance.consul-server[1].public_ip}\",\"${aws_
instance.consul-server[2].public_ip}\"]",
      "client_addr = \"${each.value.private_ip}\"",
      "ca_file =\"/etc/consul/consul.d/consul-agent-ca.pem\"",
      "verify_incoming = true",
      "verify_outgoing = true",
      "verify_server_hostname = true",
      "auto_encrypt = {",
```

```
      "  tls = true",
      "}",
      "EOF",
      "sudo mv /tmp/consul-client.hcl /etc/consul/consul.d/
consul-client.hcl",
      "sudo mv /tmp/httpd.json /etc/consul/consul.d/httpd.
json",
      "nohup python3 -m http.server 8080 &",
      "sleep 60"
    ]
  }
  provisioner "remote-exec" {
    inline = [
      "sudo systemctl start consul",
    ]
  }
  connection {
    type        = "ssh"
    user        = "ubuntu"
    private_key = local_file.private_key.content
    // private_key = file("~/hashicorp/aws/rjackson-east2.pem")
    host        = each.value.public_ip
  }
}
```

In addition to adding the CA file to the configuration, we're also adding a few other parameters:

```
      "verify_incoming = true",
      "verify_outgoing = true",
      "verify_server_hostname = true",
      "auto_encrypt = {",
      "  tls = true",
      "}",
```

The parameters that all start with `verify` direct the client to verify the certificates on all incoming messages, outgoing messages, and the server hostname that is associated with the certificates. That last bit, `auto_encrypt`, is used to enable the clients to acquire their private and public keys from the Consul server cluster. This is a simple alternative to generating certificates for each client and distributing them. Your security team most likely has a preference, or a requirement, to utilize official certificates as opposed to those created by Consul.

Now that we've gone through the Terraform code and seen how we are configuring the clients, let's look at what we are doing to the server configuration. Remember that this is done through the `user_data` script, which is called from Terraform:

```
#!/usr/bin/env bash
cat << EOF > /etc/consul/consul.d/consul-server.hcl
server = true
bootstrap_expect = 3
retry_join = ["provider=aws tag_key=Name tag_value=${server_
name_tag}"]
bind_addr = "$(curl -s http://169.254.169.254/latest/meta-data/
local-ipv4)"
advertise_addr = "$(curl -s http://169.254.169.254/latest/meta-
data/public-ipv4)"
client_addr = "$(curl -s http://169.254.169.254/latest/meta-
data/local-ipv4)"
ui = true
verify_incoming = true
verify_outgoing = true
verify_server_hostname = true
ca_file = "/etc/consul/consul.d/consul-agent-ca.pem"
cert_file = "/etc/consul/consul.d/dc1-server-consul-${server_
number}.pem"
key_file = "/etc/consul/consul.d/dc1-server-consul-${server_
number}-key.pem"
ports {
  http = 8500,
  https = 8501
}
auto_encrypt {
  allow_tls = true
```

```
}
connect {
   enabled = true
}
EOF
sudo systemctl start consul
```

Much like the clients, we have our verification parameters set to make sure that we are validating messages being received, as well as messages being sent, with our certificate files. Those files are configured in the next few lines, but notice that, unlike the clients, we have our private and public keys configured for each server. Our server cluster must be configured and created in a secure manner before enabling it to distribute certificates to the clients using the auto_encrypt configuration. I want to call out the ports that are configured. Within this configuration, we have port 8500 configured for HTTP, which is what we used in *Chapter 2, Architecture – How Does It Work?*. However, we also have an HTTPS configuration, with port 8501 specifically for TLS communication with the Consul cluster. Therefore, if you disable HTTP (set http = -1), you will secure your Consul system by preventing any unencrypted HTTP traffic. If you want to see what it would look like for an unauthenticated web browser to try to communicate with Consul, assuming the Consul system is up, point your web browser to one of the CONSUL_HTTPS_ADDR parameters provided in the Terraform output. You'll see something similar to the following:

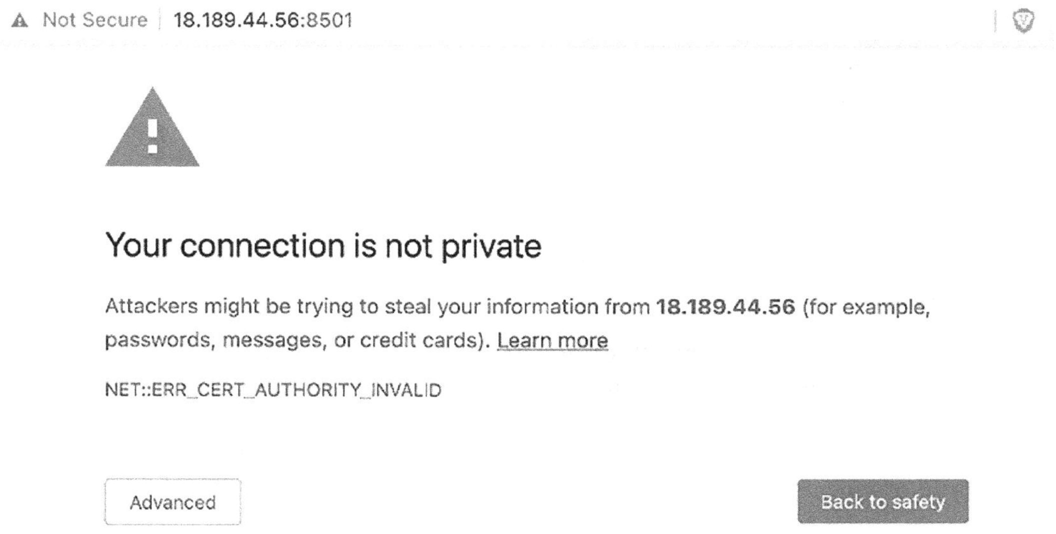

A Not Secure | **18.189.44.56**:8501 |

Your connection is not private

Attackers might be trying to steal your information from **18.189.44.56** (for example, passwords, messages, or credit cards). Learn more

NET::ERR_CERT_AUTHORITY_INVALID

Advanced   Back to safety

Figure 3.8 – Unauthorized HTTP access

If you examine the certificate, you'll see that it has been issued by the Consul CA and is not trusted. This is because the certificate hasn't been signed by an official CA; it has only been signed by Consul itself.

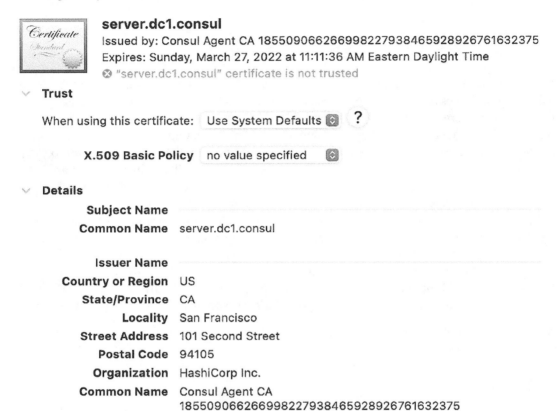

Figure 3.9 – Certificate used for HTTP communication

Thankfully, we still have port 8500 open, to facilitate our communication with the server cluster unimpaired.

But wait, there's more! Do you remember how we were saying that there were two main paths for protecting our communication? One was through mutual TLS, using certificates for the RPC communication. What about all our servers gossiping? Yup, we must protect that, too, but don't worry, that's much easier to do, and like the certificates, I've set this up for you ahead of time.

If you look in the `consul-common.hcl` file, you'll see an additional line with a simple parameter, `encrypt`. This file was placed on all the servers as part of the image build process of using Packer:

```
log_file = "/etc/consul/logs/"
log_level = "DEBUG"
encrypt = "86753098Qr2LsXLcfD+Y+S6RUuzUZsY1ZxdAj5E9c="
encrypt_verify_incoming = true
encrypt_verify_outgoing = true
```

This value isn't something I just made up – well, not entirely. By utilizing the `consul keygen` command, you can generate an encryption key that can then be used to encrypt the gossiping among your nodes. As the parameter is in the *common* configuration, all the nodes utilize the same key. We're also ensuring that each agent is encrypting their traffic (`encrypt_verify_outgoing`) and only accepting traffic that has been encrypted (`encrypt_verify_incoming`). If you want to update this encryption key (which is strongly recommended), you can use the `consul keyring` command.

At this point, you have a Consul cluster all set up, with protected messages running between all the nodes, both for RPC communication and all the gossip communication. As I stated previously, the decision to include certificates and encryption parameters as part of the build was to simplify the process. In my experience, the process of creating and distributing certificates is challenging, and even more so if you don't have officially registered domains for your machines. This process should be completed by your security team, so be sure to involve them when designing secure communication among your Consul machines.

Let's play with some ACLs!

Now that we have our cluster up and running, let's start playing with some ACLs. I didn't want to enable ACLs as part of the build process simply because without the tokens, we wouldn't be able to access our Consul system! I also wanted you to see how the system is locked down with the default deny policy, and then we would open it up slowly. We're going to be bouncing through machines quite a bit here using SSH and the key that is created as part of the build process. In a real-world scenario, like our build-out, much, if not all, of this can be performed as code with Terraform, so if you find this process painful, you might want to read a Terraform book next.

The first thing we need to do is modify our server configuration to enable ACLs. The structure of the file has already been placed on the server, so all we need to do is get that into the right place and then we can start playing. Now, we need to perform these actions on EACH server in the cluster, because if you only have ACLs enabled on one server, you might end up with dogs and cats living in harmony, and we just can't have that!

If you can still see your Terraform outputs, that's great – be sure to copy them to a temporary text file so that you can play with them. There's lots of good information there. If you're thinking, *what in the world is he talking about?*, don't worry – we can get this information by simply running `terraform output` in the same ch3 directory:

```
$ terraform output
```

The result will be similar to the following (the IP addresses will be different):

```
CONSUL_HTTPS_ADDR = "https://3.138.196.158:8501"
CONSUL_HTTP_ADDR = "http://3.138.196.158:8500"
Consul_Client_IPs = [
  "3.129.65.158",
  "3.19.228.168",
  "52.15.157.72",
  "18.216.117.182",
  "3.22.221.0",
]
Consul_Server_IPs = [
  "3.128.25.100",
  "3.138.196.158",
  "18.222.5.4",
]
```

Connect to the first server in your list using SSH and the pem key that was created for you. It will be based on your owner variable, which you set up way back at the beginning of time:

```
$ ssh -i rjackson-consul-key.pem ubuntu@3.128.25.100
```

The first thing we want to do is move that `server_acl.hcl` file to its proper location. As part of the provisioning process, this file was placed in the `/tmp` directory. Now, we need to move it to the Consul configuration directory. Once the file has been moved, we need to restart Consul. To do that, execute the following command:

```
$ sudo mv /tmp/server_acl.hcl /etc/consul/consul.d/
$ sudo systemctl restart consul
```

This needs to be done on each of the three Consul servers in your cluster. The file exists on all three servers, so you just need to SSH to each one and run the same mv command. Once you've moved the file on all three servers, select one and settle down for some fun. Find the private IP address for the server, which is your temporary home, using the `ifconfig` command:

```
$ ifconfig | grep inet
```

The output will provide you with four lines, similar to those shown here:

```
inet 192.168.100.231  netmask 255.255.255.0  broadcast
192.168.100.255
inet6 fe80::49e:7ff:fe21:bce6  prefixlen 64  scopeid 0x20<link>
inet 127.0.0.1  netmask 255.0.0.0
inet6 ::1  prefixlen 128  scopeid 0x10<host>
```

Now, you're going to export that first address as an environment variable. This will give you the ability to interact with that node on the command line. Let's move along with the current example:

```
$ export CONSUL_HTTP_ADDR=http://192.168.100.231:8500
```

Now, you can interact with the Consul CLI to play with ACLs, policies, services, and all sorts of other good stuff. To test this out, let's query the Consul members:

```
$ consul members
```

Uh oh, we got nothing. Can you guess why? Well, we've enabled ACLs, that's why! Now that you've enabled ACLs on that server, you can no longer interact with it unless you have a token. Oh no, but how do we create a token? By utilizing our good old friend, bootstrap Bill!

Whenever you start building your system with ACLs, you need to *bootstrap* the initial token. This is a very special token that provides full management abilities, kind of like your *root* user in most environments. *However, like your root user, once you've set up your ACLs, you should securely save that bootstrap token in some super-secret vault.* But for now, let's focus on getting ACLs working throughout our system. Initialize the ACL bootstrap process via the following command:

```
$ consul acl bootstrap
```

You'll see a bunch of output from this command, including a special Secret ID:

AccessorID:	aedf1ba5-5023-e0a5-643e-f1e814ec5562
SecretID:	34e87e9d-2207-f2eb-7efc-d41674eca6f0
Description:	Bootstrap Token (Global Management)
Local:	false
Create Time:	2021-04-03 22:58:49.387046985 +0000 UTC
Policies:	
00000000-0000-0000-0000-000000000001 - global-management	

This secret ID is your initial token to interact with the Consul system. As you can see, there is a single policy attached to that token for global management. Just to make things easier, we're going to utilize this token and the global management policy for all our servers. One way to do this is by exporting that token, just like we did with the IP address. Moving along with the example here, we have the following:

```
$ export CONSUL_HTTP_TOKEN=34e87e9d-2207-f2eb-7efc-d41674eca6f0
```

Now, let's run that Consul members command again. You should get the full list of servers and clients in the response.

Now that we have this deity-level token, we need to save it! Wherever you copied your IP addresses for the clients and servers, copy that bootstrap token there as well. You never know when you'll need to get this level of access. However, we do need to create something more reasonable for our servers. Again, we don't want that root token flying around, and although we can really get down to a very fine point, we're going to just set things up so that we can get used to using ACLs and learn a bit about them.

So, our first goal is going to be to create a new policy with two simple rules for our servers. We need write access to our nodes and our services. As part of this process, we need to create our policy file, which can be done using a text editor, or simply by using the following command:

```
cat > first-policy.hcl << EOF
node_prefix "" {
    policy = "write"
}
service_prefix "" {
    policy = "read"
}
EOF
```

If you were able to copy and paste correctly (or if you're using a paper book, typing properly), the file should look as follows:

```
node_prefix "" {
    policy = "write"
}
service_prefix "" {
    policy = "read"
}
```

Based on our expertise, we know that this allows the node to write all node-level operations and that it's able to read all service-level information. Using this file, we're going to create a policy within Consul. This can be done with the following command:

```
$ consul acl policy create -name "first-token" -description
"Server Token Policy" -rules @first-policy.hcl
```

The output of this command should include all the policy information:

```
ID:            c034c84d-f4f6-6bb6-4e96-8fc9a4d7dbae
Name:          first-token
Description:   Server Token Policy
Datacenters:
Rules:
node_prefix "" {
    policy = "write"
```

```
}
service_prefix "" {
    policy = "read"
}
```

Now that we have our policy, we're going to generate a new token that can then be applied to each of the server nodes. To create this token, we're going to execute the following command:

```
$ consul acl token create -description "Server Token" -policy-
name "first-token"
```

The output of that command is as follows:

```
AccessorID:       70fb05a4-58f9-ee2b-6c94-cdc52a753a58
SecretID:         5bb27adf-1401-0ccc-8bde-6c9e881e0c62
Description:      Server Token
Local:            false
Create Time:      2021-04-03 23:08:54.436055041 +0000 UTC
Policies:
    c034c84d-f4f6-6bb6-4e96-8fc9a4d7dbae - first-token
```

We now have a new `SecretID` value that we can place within the `server_acl.hcl` configuration file on each server. This will enable our server cluster to essentially manage itself. This is something you wouldn't normally want to hardcode, and a much more secure method would be to utilize the Consul secrets engine provided by HashiCorp Vault, which can create and manage those tokens for you. Again, this must be done on each server, and when complete, you must restart the server again using the `systemctl` command you used earlier. Here is an example of using the best command-line editor in the world – vim:

```
$ sudo vim /etc/consul/consul.d/server_acl.hcl
```

After configuration, the file should look similar to the following:

```
acl = {
  enabled = true
  default_policy = "deny"
  enable_token_persistence = true
  tokens {
    "default" = "5bb27adf-1401-0ccc-8bde-6c9e881e0c62"
```

```
    }
}
```

Be sure to restart Consul on each of the three servers again, which will enable all of the server nodes to now employ their new token!

Now that our server nodes have been set up, let's see what happens with our client nodes. We have five client nodes in our network, and when you run `consul members`, they are all still there. This is simply because the server is maintaining the connection with those machines. Once the Consul service on those machines is restarted, that node will no longer be able to register itself on the server cluster. If you don't believe me, try it!

Do you remember that list of IP addresses I told you to save? I hope you have it handy because we're going to need that to access the clients. Let's open a new Terminal tab, or window, and SSH to one of the clients. Any client will do, so just pick one from the list. I like to find one with special numbers if I can:

```
$ ssh -i rjackson-consul-key.pem ubuntu@3.142.52.65
```

Look at the hostname of the server. It will be right there in the command line:

```
ubuntu@ip-192-168-100-77:~$
```

On your server, run that `consul members` command again and look for that host. It should be there, but it's about to be banned from your Consul world! Restart that client agent:

```
$ sudo systemctl restart consul
```

Give it a minute or so, and then check the `consul members` output on the server again. You'll see that the agent left the cluster, but it's still there! It may be, but it is no longer authorized to join the cluster:

```
ip-192-168-100-77    3.142.52.65:8301    left    client    1.9.0
2         dc1   <default>
```

What do you think is happening on that client node? We can get a hint by checking its status. On that client node, execute the following command:

```
$ sudo systemctl status consul
```

In the output, you'll see that the agent is running, but you'll also see some errors:

```
consul.service - Consul server agent
   Loaded: loaded (/etc/systemd/system/consul.service;
disabled; vendor preset: enabled)
   Active: active (running) since Sat 2021-04-03 23:37:57 UTC;
3min 57s ago
  Process: 2990 ExecStartPre=/bin/chown -R consul:consul /var/
run/consul (code=exited, status=0/SUCCESS)
  Process: 2979 ExecStartPre=/bin/mkdir -p /var/run/consul
(code=exited, status=0/SUCCESS)
 Main PID: 2991 (consul)
    Tasks: 7 (limit: 2347)
   CGroup: /system.slice/consul.service
           └─2991 /usr/local/bin/consul agent -config-dir=/etc/
consul/consul.d/ -data-dir=/var/lib/consul/ -pid-file=/var/run/
consul/consu


Apr 03 23:39:04 ip-192-168-100-77 consul[2991]:
2021-04-03T23:39:04.935Z [DEBUG] agent.auto_config: making
AutoEncrypt.Sign RPC: addr=
Apr 03 23:39:04 ip-192-168-100-77 consul[2991]:
2021-04-03T23:39:04.943Z [ERROR] agent.auto_config:
AutoEncrypt.Sign RPC failed: addr=
Apr 03 23:39:04 ip-192-168-100-77 consul[2991]:
2021-04-03T23:39:04.944Z [ERROR] agent.auto_config: No servers
successfully responded
```

Oh; the RPC failed, and no servers successfully responded! Why do you think this happened? If you guessed ACL, then you're really starting to learn from this book. I guess I'm not that horrible at English after all!

Now that we've ostracized our client node, we need to help it get back into its family of servers. You can do this with all five clients if you like, but at this point, working with one client will be sufficient. As you may recall, part of that provisioning process included moving a `client_acl.hcl` file to the clients, similar to what we did for the servers. Let's move that client file into the proper directory:

```
$ sudo mv /tmp/client_acl.hcl /etc/consul/consul.d/
```

If we look at that file, we'll see that it is nearly identical to the server file that we put in place. There is only one thing missing – a suitable token:

```
acl = {
  enabled = true
  default_policy = "deny"
  enable_token_persistence = true
  tokens {
     "default" = ""
  }
}
```

So, before we restart that client node again, why don't we create a new policy and token for that node? This one is going to be a bit more specific. To create these items, we need to go back to the Consul server that we were using previously. If your bootstrap token is no longer in the environment, you'll need that again. You can check this by looking at the environment variables:

```
$ env | grep CONSUL
```

The output should contain the Consul address of the server that you are working on, as well as its HTTP token:

```
CONSUL_HTTP_TOKEN=34e87e9d-2207-f2eb-7efc-d41674eca6f0
CONSUL_HTTP_ADDR=http://192.168.100.146:8500
```

If you have both these values, and they are correct, then we can continue.

We're going to create a policy specifically for this node. The hostname for the *client* node I'm working with is ip-192-168-100-77. The policy rule I'm going to create is going to enable node- and agent-level write operations, but only for that node. Let's start by creating that policy file, similar to what we did previously for the servers:

```
$ cat > client-policy.hcl << EOF
node "ip-192-168-100-77" {
  policy = "write"
}
agent "ip-192-168-100-77" {
  policy = "write"
}
EOF
```

The file produced should look similar to this, but where I have my node, `ip-192-168-100-77`, you should have the hostname of your client node. Now, we have to create the policy with this file:

```
$ consul acl policy create -name "client-token" -description
"Client ip-192-168-100-77 Token Policy" -rules @client-policy.
hcl
```

The output of this command should include all the policy information:

```
ID:            a1fb3a7c-2a42-613d-4586-5dadfb422022
Name:          client-token
Description:   Client ip-192-168-100-77 Token Policy
Datacenters:
Rules:
node "ip-192-168-100-77" {
  policy = "write"
}
agent "ip-192-168-100-77" {
  policy = "write"
}
```

Just as we did previously, we must now create a new token for that policy. To create this token, we're going to execute the following command:

```
$ consul acl token create -description "ip-192-168-100-77
Token" -policy-name "client-token"
```

The output of that command is as follows:

```
AccessorID:    b0b8cee4-8efd-62d2-e22c-fc84a0dc3c92
SecretID:      776f0c07-6352-8d2f-8fc7-23f94cbafd74
Description:   ip-192-168-100-77 Token
Local:         false
Create Time:   2021-04-04 00:02:08.863705384 +0000 UTC
Policies:
   a1fb3a7c-2a42-613d-4586-5dadfb422022 - client-token
```

We now have a new `Secret ID` value that we can place within `client_acl.hcl`. Wouldn't this be so much easier if we used Terraform and Vault?

Go back to your tab that's logged in to the client that you are working with and edit the `client_acl.hcl` file to insert your new token:

```
$ sudo vim /etc/consul/consul.d/client_acl.hcl
```

This is what the file should look like – with your token of course, not mine!

```
acl = {
  enabled = true
  default_policy = "deny"
  enable_token_persistence = true
  tokens {
    "default" = "776f0c07-6352-8d2f-8fc7-23f94cbafd74"
  }
}
```

OK; we have our configuration and our token. Are you ready to restart that Consul service now? I know I am!

```
$ sudo systemctl restart consul
```

You can either check that status again on the client node or if you are more excited, go back to your server node and run another `consul members`. Do you see the node you've been working on? I certainly hope so!

As you may recall, all those nodes are running both the Consul service as well as an HTTPD service that was deployed when we set up the system. Let's check to make sure those services are registered on that node. We can do this via the command line, on the server node, using the following command:

```
$ consul catalog services -node=ip-192-168-100-77
```

> **Note**
>
> If you run a command like this and end up with an ACL error, you might not be on the server node that has your bootstrap token in the environment variable.

Of course, for the node, enter the name of the client node that you've been working with. I'm not expecting my node to be living on your system, but look what happens with the output of that command:

```
No services match the given query - try expanding your search.
```

Why did this happen? If you're confused, look at the policy that we created for that client token:

```
$ consul acl policy read -name "client-token"
```

For the rules that are associated with that token, we have a node-type rule and an agent-type rule, but no service-type rule! Well, of course, none of our services will be registered if we don't have the services permitted to write to the catalog! We can fix this by modifying the policy on the server node, without even touching the client. This workflow can be very useful when you have several clients and you don't want to redistribute new tokens.

Edit the `client-policy.hcl` file and add a rule that allows write access to the `httpd` service:

```
$ sudo vim ~/client_acl.hcl
```

Here are the additions to that file:

```
service "httpd" {
  policy = "write"
}
```

When complete, the entire file should look like this:

```
node "ip-192-168-100-77" {
  policy = "write"
}
agent "ip-192-168-100-77" {
  policy = "write"
}
service "httpd" {
  policy = "write"
}
```

This is very specific to our node, as well as to the two services that are running on our client node – just as things should be.

Let's update our client-token policy with the new rule:

```
$ consul acl policy update -name "client-token" -rules=@client-
policy.hcl
```

The output should confirm that your new service rules have been written to the policy:

```
ID:            a1fb3a7c-2a42-613d-4586-5dadfb422022
Name:          client-token
Description:   Client ip-192-168-100-77 Token Policy
Datacenters:
Rules:
node "ip-192-168-100-77" {
  policy = "write"
}
agent "ip-192-168-100-77" {
  policy = "write"
}
service "httpd" {
  policy = "write"
}
```

Great! We've updated our policy on our server node, and we never had to return to the client. Here's the final test; does your httpd service now show up in your catalog for that node? Run the following command to find out:

```
$ consul catalog services -node=ip-192-168-100-77
```

You should see the httpd service listed in the output. If you do, stand up and scream to the world, *I have mastered Consul Access Control Lists!* Well, maybe *mastered* is a bit of an exaggeration, but you should be very proud of what you've been able to accomplish here!

Summary

Wow – another long chapter! I'm not sure about you, but I am seriously wiped out! Although this is a *getting started* book, the concept of secure communication using TLS certificates is not a light topic for the family dinner. As Consul is such a critical service to the network, though, with the application and service information it maintains, the necessity to authenticate your nodes and encrypt their communication paths can't be understated. Furthermore, Consul ACLs provide such a level of flexibility that they can be very overwhelming. However, their power and control should be a pinnacle component in any production environment. I would strongly encourage you to continue playing with the other client nodes, create some new policies and tokens for them, and see what you can manage. If you're worried about killing something, don't be. You can always use Terraform to destroy and rebuild the infrastructure. On that topic, don't forget to destroy what you've built today… we don't want to end up with an AWS bill at the end of the month to struggle with! Once you've destroyed your infrastructure, kick your feet up and have a beverage. The next chapter will be a bit lighter (I promise). There, we'll be discussing how to extend Consul across multiple data centers and connecting clusters. We're almost halfway there!

4

Data Center (Not Trade) Federation

I'm very happy to tell you (and myself) that after that very challenging chapter dealing with Consul security aspects, our discussion about extending your Consul cluster will be much lighter. In fact, as many of the features we'll be discussing in this section are specific to Consul Enterprise, there won't be any exercises!

In the previous chapters, we focused primarily on setting up and securing our Consul clusters. However, as Consul is implemented in enterprise environments, we want to look at how to scale the Consul system itself and extend that Consul functionality across multiple data centers or regions. To understand the expansion of Consul, we'll be reviewing the following topics:

- Scales – Nope, we won't be discussing reptilian or musical scales, but rather how to scale the Consul system. We've seen in earlier chapters how increasing server clusters can negatively impact the time to achieve consensus. So, how can we ensure that we have enough server structure to manage the cluster as we increase our client count? How do we manage upgrades across such a vast system?

- Subdivisions – After we look at how to scale Consul into different areas and manage that increased scale, we're going to look at how to make best use of that expanded system. The primary result of this operation is to offer autonomous Consul systems within an enterprise environment without giving up centralized control.

- Federation – In this global world, services rarely reside in one region or data center. In this section, we're going to see how we connect server clusters across multiple regions to ensure that our services can communicate regardless of their location.

Although we won't be applying the knowledge gained in this chapter directly, I will be including code samples to help further explain the implementation of the concepts. So, with that, let's start playing our scales!

Protection and scale

Although Consul agents can work with services on external nodes, best practices (and best flexibility) recommend having a Consul agent co-resident on the nodes running the connected services. In many enterprises, Consul is connecting thousands of services (or more), which means Consul clusters of thousands, or tens of thousands, of nodes. Certainly, we don't want only three to five server nodes managing a cluster of that size, so let's start with expanding the Consul server cluster.

As we learned in *Chapter 2, Architecture – How Does It Work?*, bigger is not always better when it comes to the size of your consensus pool. We need at least three servers to obtain a quorum, but as that server count increases, it's going to take much longer for the pool to elect a leader. In order to resolve this, Consul offers the ability to configure Consul server agents as a *read replica*. When set with this flag, the agent doesn't participate in any quorum election, but it still helps deal with node and service registration, just like any other server within the cluster.

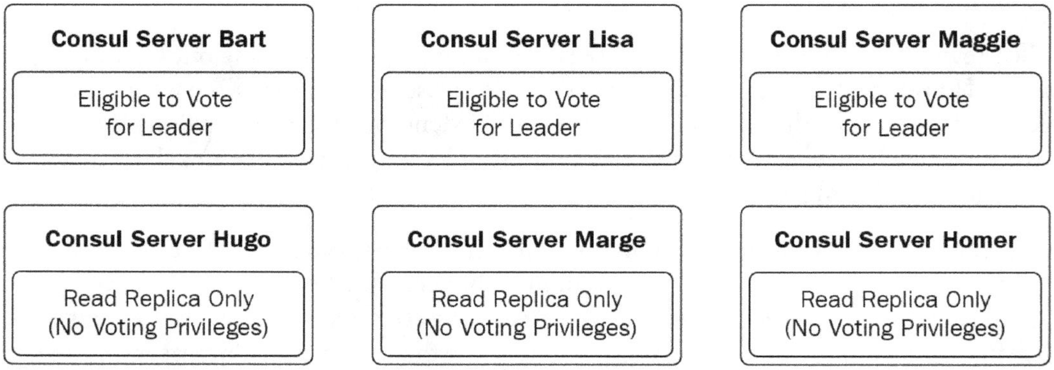

Figure 4.1 – Read replica servers

Using the preceding example, let's look at our server cluster again. We started with Bart, Lisa, and Maggie as our three primary Consul server nodes. We're able to attain consensus quickly, but as Springfield grows, we need to start extending our Consul network with more services. So, we end up adding three servers into the cluster: Hugo, Marge, and Homer. In order to ensure that neither Hugo, Marge, nor Homer has the ability to vote for a server cluster leader, they are designated as *read replicas* via a Boolean parameter in the configuration:

```
server = true
bootstrap_expect = 3
retry_join = ["provider=aws tag_key=Name tag_value=${server_
name_tag}"]
bind_addr = "$(curl -s http://169.254.169.254/latest/meta-data/
local-ipv4)"
advertise_addr = "$(curl -s http://169.254.169.254/latest/meta-
data/public-ipv4)"
client_addr = "$(curl -s http://169.254.169.254/latest/meta-
data/local-ipv4)"
ui = true
read_replica = false
```

When we set this value to `true`, Consul considers that node to be a *read replica* node, and reflects that configuration in the output of the `consul operator raft list-peers` command:

Node	ID	Address	State	Voter	RaftProtocol
ip-192-168-100-213	bed9a1c9-300a-93fd-3fc7-874056f8a79c	3.142.189.139:8300	leader	true	3
ip-192-168-100-187	76bd2284-7060-fda3-b873-b168247f4385	3.21.19.228:8300	follower	true	3
ip-192-168-100-66	63b03028-f8db-dc2f-00ba-c6730f1bed72	3.128.179.103:8300	follower	true	3
ip-192-168-100-40	8521bb8f-21d5-e1cb-cd85-68d57758857b	3.131.95.33:8300	follower	false	3
ip-192-168-100-27	86b14590-c26c-06da-6989-f8de786e32ec	3.143.222.182:8300	follower	false	3
ip-192-168-100-157	397eecab-05e5-aa3a-48cc-6acc29d23134	18.191.133.108:8300	follower	false	3

Figure 4.2 – List of raft peers

Now, we have six servers in our cluster, and, yes, this is a completely logical number of servers for a server cluster. Actually, it can be quite ideal if you want to offer Consul services across three unique availability zones (it's always good to protect against any sort of geographical failures). Let's consider a situation where we have these six server agents spread across three availability zones, as shown in the following figure:

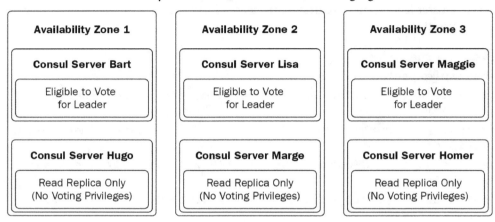

Figure 4.3 – Servers deployed to availability zones

What shall occur when we lose one of those server nodes? If Bart goes away, we will still have plenty of nodes to maintain a quorum, but we only have two server agents that are actually participating in the leadership election. Furthermore, now that Bart has left zone 1, we no longer have a leader within that zone. Leader election has been left to Maggie and Lisa. Neither Hugo, Marge, nor Homer can participate in that election, as they are configured to be *read replicas*. The cluster can continue to operate, but if anything happens to either zone 2 or 3, our cluster will fail to have a leader. If only there was some way to associate Consul servers with zones, such as what is available in the cloud infrastructures. We can call them **redundancy zones**!

This concept of redundancy zones can extend far beyond the case of a failure of a node. For example, as we are scaling our server cluster to support the thousands of nodes we're deploying, we need to ensure that we maintain the proper number of servers capable of electing a leader. Recall that there is a sweet spot here for a Consul cluster. If we don't have enough servers, we are vulnerable to failure. If we have too many servers, it can take too long to elect a leader. This might be manageable in a steady-state system, but if Heraclitus taught us anything, it's that the only thing constant in this world is change. Simply the act of performing upgrades, either to the Consul agent itself or the environment in which Consul lives, can completely alter our pool of electable servers and disrupt our cluster. With Consul being such a critical part of the communication system, we must protect against these situations! The protection comes in the form of **Consul Autopilot**, but don't worry – this autopilot isn't inflatable or sentient.

Consul Autopilot

Consul Autopilot itself is an open source feature; however, it largely operates in the background. As a matter of fact, you've already been using Autopilot! Much of the Consul server functionality with respect to creating and maintaining the quorum is performed by Autopilot. Let's go through the configuration parameters of Autopilot to understand why it's such a useful little engine.

To see what configuration exists by default within your cluster, just run the following command:

```
# consul operator autopilot get-config
```

Assuming you're on the same Consul version, the output should be identical to the following, as these are the default parameters:

```
CleanupDeadServers = true
LastContactThreshold = 200ms
MaxTrailingLogs = 250
MinQuorum = 0
ServerStabilizationTime = 10s
RedundancyZoneTag = ""
DisableUpgradeMigration = false
UpgradeVersionTag = ""
```

In most cases, these default parameters will be fine, but if you need or want to change them, it would be good to understand what they control:

- CleanupDeadServers – In the inevitable case of a server failure, you're going to want to replace that failed server as quickly as possible. This can be done through a manual process or some automated method. However, within the Consul member list, a failed server could stick around for a while if it didn't say goodbye before it left. This could cause some confusion around the quorum of servers and the leader election. This value set to true enables Consul to remove any failed servers from the server cluster as soon as a new one is available to take over.

- LastContactThreshold – Much like the name indicates, if the leader hasn't heard from a particular server node within this time period, the leader will assume the node is failed.

- `MaxTrailingLogs` – Recall that the data is synchronized among the Consul server nodes via logs that indicate service catalog status along with other useful information. The leader maintains these logs and distributes them to peer nodes, but what if a node falls too far behind? This parameter indicates just how far behind a server can be from the leader, defined by the number of log entries (not logs) before the node is considered to be unhealthy.

- `MinQuorum` – This value defines the minimum number of servers that need to be in a cluster before Autopilot can start pruning servers that have been marked as dead or unhealthy. This corresponds directly to the parameter to clean up dead servers.

- `ServerStabilizationTime` – We all need some time to be considered a stable and valuable member of society. My therapist is still trying to figure out the time I need; however, our cluster server nodes are much more predictable than the human brain. This value indicates how long the cluster leader should wait before deeming a new member of the cluster to be healthy.

To these five parameters, three have been added for an enterprise system:

- `RedundancyZoneTag` – Each server can be configured with metadata to help with any number of actions. One of those actions is to define what redundancy zone that server resides within. This parameter indicates which tag is used for that designation.

- `DisableUpgradeMigration` – This Boolean parameter directly impacts how Consul performs throughout the process of an upgrade. We won't be talking about upgrading Consul within this book, however; upgrading any distributed system such as Consul should consist of a rolling update of individual nodes. This eliminates the need to take down the entire cluster, upgrade all nodes, and bring them all back at once (something that would cause a pretty significant service outage). With Consul, when we start upgrading servers, as soon as enough servers are on the new version to support a quorum, Autopilot would start promoting them to the voters. If we set this value to `true`, Autopilot would not automatically promote the new servers as voters.

- `UpgradeVersionTag` – In the event that you want to manually override a version of a Consul agent, perhaps when shifting configuration parameters across a cluster, this parameter defines what metadata would be used to indicate the *version* of the server. Without this value, Autopilot simply looks at the version being reported by the Consul software itself.

We aren't really going to be playing with any of these parameters, with the exception of `RedundancyZoneTag`. This will also help us understand the use of the metadata within the Consul agent configuration. So, let's look at a new configuration for our server nodes with the metadata and redundancy zone configured:

Consul server configuration with redundancy zones

```
server = true
bootstrap_expect = 3
retry_join = ["provider=aws tag_key=Name tag_value=rjackson-
consul-server-instance"]
bind_addr = "192.168.100.20"
advertise_addr = "3.17.154.23"
client_addr = "192.168.100.20"
ui = true
read_replica = false
node_meta {
   stooge = "moe"
}
autopilot {
   redundancy_zone_tag = "stooge"
}
```

Here, we have two new configuration blocks, one for node_meta and another for autopilot. Not a lot has been entered into these blocks, but this is all that is necessary to set up the redundancy zones. First, I defined some sort of metadata for each node. I'm calling the key for that metadata stooge, and the value is either moe, larry, or curly, indicating how I'm grouping the servers. It might be good for your metadata to indicate the actual zone in accordance with your own network structure, but I wanted to use the stooge parameter for fun and to show that these keys can really be just about anything. We can see the metadata configured for each node within the UI, as shown in the following figure:

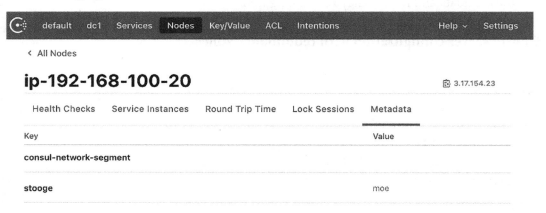

Figure 4.4 – Consul agent metadata

You can also filter the output of the CLI based on the metadata:

```
# consul catalog nodes -node-meta=stooge=moe
Node                ID          Address        DC
ip-192-168-100-20   8a1da666    3.17.154.23    dc1
ip-192-168-100-74   d80b81de    3.135.205.26   dc1
```

Note that in these configurations, the *read replica* flag is set to false for all servers. When we designate a server as a *read replica*, we are essentially telling Consul that we will manage the existence of *read replicas* and we don't need Autopilot to manage that for us. When we utilize redundancy zones, Autopilot ensures that there is at least one voting member in each redundancy zone. Any additional servers that join the particular zone will automatically be recognized as *read replicas*. To see this in action, I want to provide a new diagram with the appropriate server names and the associated metadata from my test system:

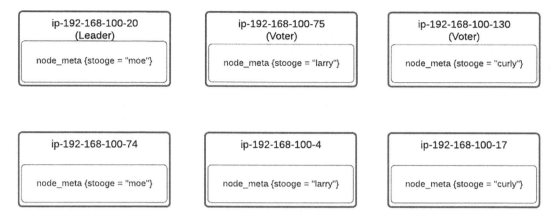

Figure 4.5 – Consul redundancy zone setup

Along the top row in this diagram, we have all of our voting members, with the first node elected as the leader. We can also see the metadata configured for each node. We can also see the voter status based on the `consul operator raft list-peers` command:

```
Node                ID                                    Address               State     Voter  RaftProtocol
ip-192-168-100-130  b48c7778-bf22-d37a-d11a-54b4f6368342  18.222.179.118:8300   follower  true   3
ip-192-168-100-75   e211c429-9b12-00cf-f15a-2f03a13cdbd2  3.139.238.20:8300     follower  true   3
ip-192-168-100-20   8a1da666-1df7-74ed-64f7-536017fae610  3.17.154.23:8300      leader    true   3
ip-192-168-100-74   d80b81de-a4cb-ba2e-d23c-fc7d2037163f  3.135.205.26:8300     follower  false  3
ip-192-168-100-17   43c87a94-4eb0-0ac2-da4d-17c4a9532f6a  3.142.173.170:8300    follower  false  3
ip-192-168-100-4    0babb7ce-a1ed-8be2-7ad4-d0182cc2e0a0  3.16.139.57:8300      follower  false  3
```

Figure 4.6 – Raft voter status

On the Consul leader node, `ip-192-168-100-20`, I'm going to shut down the network interface, causing it to lose all communication with the other nodes in the cluster, using the `sudo ifconfig eth0 down` command.

> **Important Note**
> If you're trying this on your own, you're going to lose connection to the server and will need to restart it via the AWS Management Console.

If I do this right, node `ip-192-168-100-74` will be converted from a non-voter value (`voter = false`) to a voter value within the cluster. Our redundancy zones are configured to match based on the `stooge` metadata key, and both `ip-192-168-100-20` and `ip-192-168-100-74` are of `stooge` type `moe`:

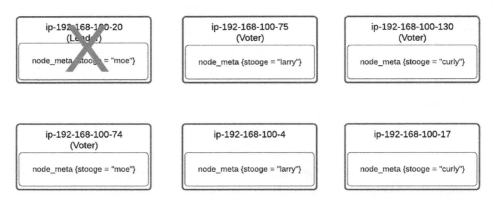

Figure 4.7 – Redundancy zone autopilot adjustment

This result can again be viewed with the `consul operator raft list-peers` command:

```
Node                     ID                                    Address               State     Voter  RaftProtocol
ip-192-168-100-130       b48c7778-bf22-d37a-d11a-54b4f6368342   18.222.179.118:8300   leader    true   3
ip-192-168-100-75        e211c429-9b12-00cf-f15a-2f03a13cdbd2   3.139.238.20:8300     follower  false  3
ip-192-168-100-74        d80b81de-a4cb-ba2e-d23c-fc7d2037163f   3.135.205.26:8300     follower  true   3
ip-192-168-100-4         0babb7ce-a1ed-8be2-7ad4-d0182cc2e0a0   3.16.139.57:8300      follower  true   3
ip-192-168-100-17        43c87a94-4eb0-0ac2-da4d-17c4a9532f6a   3.142.173.170:8300    follower  false  3
```

Figure 4.8 – Down to five raft peers

Notice that the promotion of `ip-192-168-100-74` wasn't the only adjustment to the assignments of voters. When Autopilot engages and readjusts the cluster, the main goal is to ensure that we have one voter in each redundancy zone.

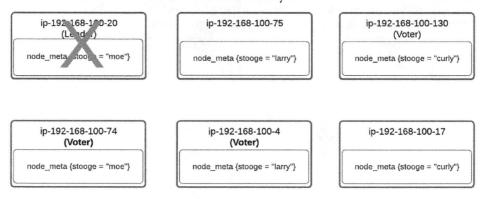

Figure 4.9 – Results of Autopilot

Throughout this section, we've seen how we can expand the Consul server cluster in a scalable way, but we've also learned how Autopilot helps maintain some level of sanity within the cluster itself. Now that we've perfected the ability to scale our Consul cluster effectively, we can start exposing the beauty of Consul to all of our colleagues! After all, once we find and learn how to use a product so powerful and helpful, don't we want to share that information with others?

Subdivisions

Utilizing the **Consul namespace** feature, we are able to effectively subdivide our entire Consul system into multiple mini-systems, each with its own autonomy. This offers multiple benefits to the greater enterprise organization:

- It utilizes a single centralized team to operate and manage Consul at a corporate scale. This not only provides consistency with how services are managed but also reduces the overhead of managing multiple independent Consul systems.

- It enables individual teams to manage and secure their own applications without relying on multilayer management and control. Although the centralized operators of Consul have full control, autonomy can be granted to teams (within bounds) to manage their services as they see fit.

- Splitting up Consul not only subdivides the operations of the system but also the service catalogs themselves. This enables organizations to create services that are completely isolated from other aspects of the Consul catalog. This not only enhances the security posture of service management but also prevents groups with duplicate service names from stepping on each other.

Essentially, by applying **namespaces** across our Consul cluster, we are able to layer a level of abstraction above the nodes to enable independent service management. This doesn't separate the Consul nodes within the system but rather provides a way to isolate the services running on that system. Let's think of Consul supporting types of trees. There are two main classifications for trees to consider, conifer and deciduous. If you aren't an arborist, don't worry – just note that conifers are typically evergreen trees (such as cedar, cypress, pine, and fir), while deciduous trees typically lose their leaves in the fall and then seem to come back to life in the spring (such as maple, oak, acacia, and birch). These classifications represent our namespaces, while the types of trees represent the service.

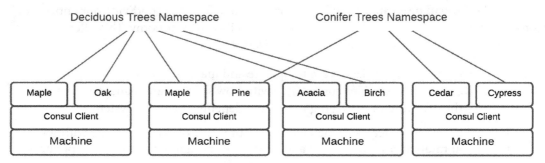

Figure 4.10 – Segmenting tree types with namespaces

To demonstrate the concept, I'm going to create two namespaces within my system. Each namespace has its own definition file, which at this point is very basic, including only the name and description of the namespace:

> **Important Note**
> Note that namespaces can only be configured and managed through either the Consul CLI or API.

shawn-namespace.hcl

```
name = "shawn"
description = "Shawns namespace for his funny antics and vision
```

gus-namespace.hcl

```
name = "gus"
description = "Gus' namespace to track his pharmaceutical
distribution"
```

Creating the namespaces themselves only requires a simple `write` command. For good measure, we're going to view the namespaces we created as well:

```
# consul namespace write shawn.hcl
Name: shawn
Description:
   Shawns namespace for his funny antics and visions
# consul namespace write gus.hcl
Name: gus
Description:
   Gus' namespace to track his pharmaceutical distribution
# consul namespace list
default:
   Description:
      Builtin Default Namespace
gus:
   Description:
      Gus' namespace to track his pharmaceutical distribution
shawn:
   Description:
      Shawns namespace for his funny antics and visions
```

Note that there is a default `namespace` value that will capture all the data that hasn't been explicitly associated with a namespace.

Now that the namespaces have been created, we can see them within the Consul UI:

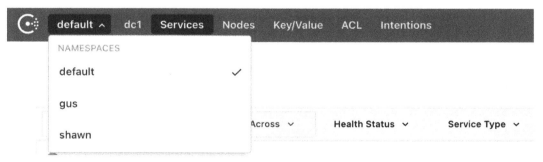

Figure 4.11 – Namespace selection within the Consul UI

If we want to register a service with a specific namespace (not the default namespace), all we need to do is specify the desired namespace within the configuration of the service itself. In previous sections, we observed an httpd service that Consul was managing. I've modified that JSON file slightly, just to indicate that this service will be running within the gus namespace.

httpd.json service definition

```json
{
    "service": {
      "name": "httpd",
      "namespace": "gus",
      "tags": [
        "python-server"
      ],
      "port": 8080,
      "check": {
        "args": [
          "curl",
          "localhost:8080"
        ],
        "interval": "10s"
      }
    }
}
```

When we have our services configured and designated to a particular namespace, as we've done here with the httpd service, then only members of that namespace can interact and obtain the information associated with that service. To follow on from my example, if I query the services that are associated with my Consul system, I only end up with one service:

```
$ consul catalog services
consul
```

Oh no! Isn't my `httpd` service running? I can assure you that the service is running, only in a separate and distinct `namespace`:

```
$ consul catalog services -namespace gus
 httpd
```

Now, being able to separate these services can help greatly; however, if we aren't securing those namespaces uniquely, then we really aren't getting the true benefit of these subdivisions. If those nerves are acting up from the memories of our ACL work, you need to trust your intuition. When we start creating namespaces within our Consul system, it's kind of like developing a management hierarchy.

If you recall, when we were playing with ACLs in order to bootstrap the system, we needed to create a bootstrap ACL token that gave us full administrative privileges over the entire Consul system. Using a grossly oversimplified analogy, we can view this administrator, or administrative group, as the governor of the Consul system. We'll call him **Governor Lepetomane**. Based on his credentials, represented by a unique token, he is able to set the rules for his jurisdiction that all of the towns and communities within must abide by. However, each town still maintains some level of autonomy. **Mayor Johnson**, with his own credentials (token), is able to create his own rules and regulations for the town of Rock Ridge, so long as those rules comply with those set forth by Governor Lepetomane. The management of the Consul namespace structure is nearly identical.

When we bootstrap the ACL functionality within Consul, we create a token that abides by the default global management policy. Using this token, we are able to create new policies, and corresponding tokens, to enable very fine-grained management of everything Consul can do. When we create our namespaces, we need to give those operators their own management policies, but we can't have them exercising their authority over neighboring namespaces. Therefore, each namespace gets its own *operator*-level token that the administrators of those namespaces can use to create their own policies, rules, services, and so on. So how are these two default policies different?

> **Important Note**
> Namespaces allow us to subdivide services and policy; however, Consul cluster nodes are visible to all namespaces.

Let's start by creating a *namespace-management*-level policy for Shawn's `namespace`:

```
consul acl token create -namespace shawn -description "Shawn's
Management Token" -policy-name "namespace-management"
```

Now we have a default operational-level token assigned to the `shawn` namespace. Utilizing the Consul CLI, we can observe the existence of both policies. Note that in order to see the *namespace-management* policy, I need to include the flag for the appropriate namespace:

```
$ consul acl policy list
global-management:
    ID:            00000000-0000-0000-0000-000000000001
    Namespace:     default
    Description:   Builtin Policy that grants unlimited access
    Datacenters:


$ consul acl policy list -namespace=shawn
namespace-management:
    ID:            2226f072-68ae-9082-6d8d-7c9bdb1a987c
    Namespace:     shawn
    Description:   Default policy that grants all available
privileges within a single namespace
    Datacenters:
```

Utilizing the `consul acl policy read` command, I'm able to look at the details of those two policies. Again, for the `shawn` namespace, I need to include the `namespace` flag in the command. *Figure 4.12* provides a side-by-side view of the policy differences:

ID: 00000000-0000-0000-0000-000000000001 Name: **global-management** Namespace: **default** Description: Builtin Policy that grants unlimited access Datacenters: Rules:	ID: 2226f072-68ae-9082-6d8d-7c9bdb1a987c Name: **namespace-management** Namespace: **shawn** Description: Default policy that grants all available privileges within a single namespace Datacenters:
acl = "write"	acl = "write"

agent_prefix "" {
 policy = "write"
}
event_prefix "" {
 policy = "write"
}	

key_prefix "" {
 policy = "write"
}
	key_prefix "" {
	 policy = "write"
	}

keyring = "write"	

node_prefix "" {
 policy = "write"
}
	node_prefix "" {
	 # node policy is restricted to read
	within a namespace
	 policy = "read"
	}

operator = "write"
query_prefix "" {
 policy = "write"
}	

service_prefix "" {
 policy = "write"
 intentions = "write"
}
session_prefix "" {
 policy = "write"
}
	service_prefix "" {
	 policy = "write"
	 intentions = "write"
	}
	session_prefix "" {
	 policy = "write"
	}

namespace_prefix "" {
    acl = "write"
    key_prefix "" {
      policy = "write"
    }
    node_prefix "" {
      policy = "read"
    }
    session_prefix "" {
      policy = "write"
    }
    service_prefix "" {
      policy = "write"
      intentions = "write"
    }
}	

Figure 4.12 – Bootstrap and namespace policies

Notice that there is an entire section of the bootstrap policy that is directly related to namespace management. Therefore, those with this very privileged policy are able to manage ACLs, keys, nodes, sessions, and services within each namespace. In other words, our governor has the ability to manage and control items at the town level, but mayors within the town have no authority over other towns. Additionally, although at the bootstrap level we have the ability to write to nodes within the cluster, at a namespace level, as the namespace has access to all nodes, it can only read.

As we've seen, namespaces are able to segment our entire Consul system into individual mini-systems, providing autonomy to the operators of those systems. However, in some highly secure environments, it may be required to actually divide Consul at the network layer, not just the application layer.

Our entire Consul cluster, every node in every namespace, communicates utilizing that gossip protocol we talked about so very long ago. We're not going to delve too deep into this topic, as it is quite advanced. However, consider the situation where we might have thousands of Consul nodes distributed throughout our data center network, a situation not uncommon in production environments. If you remember, when we started the discussion about namespaces with the tree example (*Figure 4.10*), the separation occurred one layer above the Consul nodes. When we segment the Consul network, we are actually dividing up those gossip pools into smaller pools, each communicating on their own network ports.

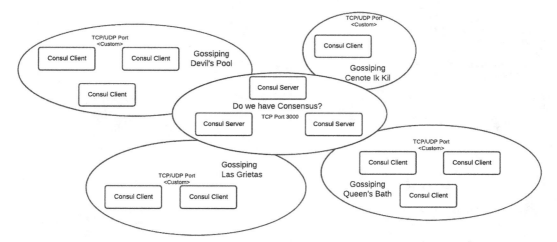

Figure 4.13 – Reducing gossip pools with network segmentation

I've tried to illustrate this segmentation in *Figure 4.13*, using different types of pools. In this example, all network segments, or *pools*, are able to communicate with the core centralized server cluster. However, due to firewall or other network-based divisions, the pools are not able to communicate with each other. By utilizing Consul network segmentation, we can still manage the overall Consul cluster, services, and namespaces within the divided network.

Now that we've successfully conquered an entire data center with Consul services, it's time to continue with the Brain's goal of taking over the world. So come along Pinky, let's start discussing how to connect our Consul clusters.

Connecting clusters

To think that all of our services are going to reside within a single data center might have worked 20 years ago when I started writing this book (so it seems). The world has truly evolved to a continuously connected mobile environment, although some might argue that this was actually devolution. Regardless, the only way to really maintain the communications among our services in this world is to ensure that their availability and location can be discovered from multiple geographic areas. Whether I'm in Boston, Berlin, or Melbourne, the applications and services I've grown to rely on must be available! All that we have done within this chapter has led us to this point!

Throughout the book, we've been talking a lot about *gossiping* and *pools*, and we're going to continue that discussion. However, as you can imagine, the amount of data that we have communicated within a single Consul cluster can be quite extensive, especially as we onboard all of our services. If we wanted, we could simply start deploying Consul nodes into different data centers and regions and be done with that. Unfortunately, this would cause a great deal of traffic to be broadcast across the entire network. Furthermore, spreading Consul server nodes across data centers can impact the availability of your server cluster, even with the brilliant Autopilot.

When you join Consul clusters across a **Wide Area Network** (**WAN**), the clusters don't share catalog information. Each server cluster only becomes aware of the other, so when a request for a service is made, that request can be transmitted to another data center to check for service availability.

So far, we've only been working in our default data center, imaginatively named *dc1*. For this example, I'm going to be creating two data centers, **Jake** and **Elwood**. The data center for the particular Consul agent is defined within the configuration with a simple `datacenter` parameter. Take this, for example:

```
server = true
datacenter = "Elwood"
bootstrap_expect = 3
```

I've set up a system with three server nodes in each of the two data centers, Jake and Elwood. Joining the two data centers is pretty simple within the configuration as well as via the command line. Within the configuration, we utilize a `retry_join_wan` command, similar to the `retry_join` command we already use for the Consul nodes to discover each other. Additionally, you only require a connection between one server in each data center, as they then learn about each other. However, I would strongly recommend including multiple servers in the event that one is not available when the two data centers need to join up:

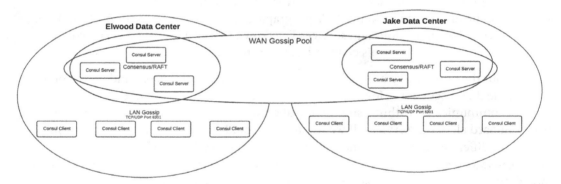

Figure 4.14 – Jake and Elwood getting the band together

Once the data centers have been joined, you can get a view of all of the servers using the `consul members -wan` command:

```
Node                        Address              Status  Type    Build      Protocol  DC       Segment
ip-192-168-100-140.elwood   3.133.92.242:8302    alive   server  1.9.4+ent  2         elwood   <all>
ip-192-168-100-180.elwood   18.216.77.187:8302   alive   server  1.9.4+ent  2         elwood   <all>
ip-192-168-100-203.jake     18.188.194.17:8302   alive   server  1.9.4+ent  2         jake     <all>
ip-192-168-100-60.jake      3.140.201.212:8302   alive   server  1.9.4+ent  2         jake     <all>
ip-192-168-100-81.elwood    3.140.252.128:8302   alive   server  1.9.4+ent  2         elwood   <all>
ip-192-168-100-97.jake      3.135.63.108:8302    alive   server  1.9.4+ent  2         jake     <all>
```

Figure 4.15 – Clusters banded together

As stated previously, the service information isn't carried across the data centers. What this does allow is the ability to search for services in data centers outside of your own; for example, here I'm working in the jake data center:

```
$ consul catalog services
consul
```

Only consul is showing up as a registered service. However, here I'm working in the elwood data center:

```
$ consul catalog services
consul
httpd
```

We can see that the httpd service does show up. If we want to query other data centers, all we need to do is add a flag to the query. This can be done in both the CLI as well as via the API:

```
$ consul catalog services -datacenter elwood
consul
httpd
```

To query the service catalog in the Elwood data center using the API, the following curl command provides an example:

```
$ curl "http://3.140.201.212:8500/v1/catalog/
services?dc=elwood" | jq

{
  "consul": [],
  "httpd": [
    "python-server"
  ]
}
```

Notice in that API call that I'm actually calling the IP address for a Consul server in the jake data center. Within my test environment, this is very easy because it is all on one flat network. Within a production environment, network connectivity for both the addresses and ports between the Consul servers is obviously required. This is not always so easy to ensure when working across data centers, especially those in different cloud environments. Furthermore, when you start connecting more data centers, they all start participating in the WAN gossip protocol. This can make things fairly meshy because every Consul data center is sharing data with every other Consul data center. Essentially, we are creating broadcast-level communications between each data center.

In many enterprise environments, it is preferred to have more direct control over which data centers are communicating. This enables the Consul administrators to design more point-to-point or hub-and-spoke-type architectures federating multiple Consul clusters. Instead of gossiping, or broadcasting the data, we are creating **remote procedure call (RPC)** links between the data centers.

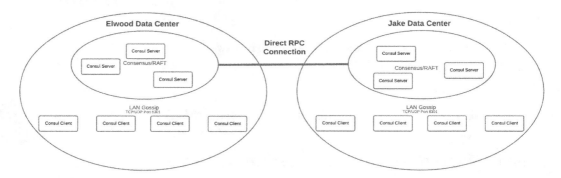

Figure 4.16 – Federating data centers via RPC

To accomplish this, the first thing we're going to do is create our network areas in each data center, with the other data center as the peer.

Here it is from the Elwood data center:

```
consul operator area create -peer-datacenter=jake
```

Here it is from the Jake data center:

```
consul operator area create -peer-datacenter=elwood
```

This allows us to create those network areas within our *own* data centers in preparation for the join. Each of those network areas is identified with an area ID, which is now visible with the `consul operator area members` command:

```
Area                                    Node                             Address              Status   Build      Protocol   DC        RTT
e294c42b-efce-4c3b-796b-0de93fc51d1f     ip-192-168-100-140.elwood        3.133.92.242:8300    alive    1.9.4+ent  2          elwood    3.911967ms
e294c42b-efce-4c3b-796b-0de93fc51d1f     ip-192-168-100-180.elwood        18.216.77.187:8300   alive    1.9.4+ent  2          elwood    0s
e294c42b-efce-4c3b-796b-0de93fc51d1f     ip-192-168-100-81.elwood         3.140.252.128:8300   alive    1.9.4+ent  2          elwood    3.6866ms
```

Figure 4.17 – Area ID identification for cluster members

> **Important Note**
> This command is only available on **Consul Enterprise** systems.

In addition to the area identifiers, and the data center of course, we can also see the **round-trip time** (**RTT**) between the various servers. Based on this output, you can see I was on the `ip-192-168-100-180` server when I ran this command. Now I'm ready to join the two data centers. From the Elwood data center, I execute the following command:

```
consul operator area join -peer-datacenter=jake 18.188.194.17
```

I'm only using one IP address from the Jake data center, simply because once the join occurs, information about the peer members in the data center will be shared. And now, when I run the `consul operator area members` command, I can see all of the members in the data centers joined.

```
Area                                    Node                             Address              Status   Build      Protocol   DC        RTT
e294c42b-efce-4c3b-796b-0de93fc51d1f     ip-192-168-100-140.elwood        3.133.92.242:8300    alive    1.9.4+ent  2          elwood    690.177µs
e294c42b-efce-4c3b-796b-0de93fc51d1f     ip-192-168-100-180.elwood        18.216.77.187:8300   alive    1.9.4+ent  2          elwood    0s
e294c42b-efce-4c3b-796b-0de93fc51d1f     ip-192-168-100-81.elwood         3.140.252.128:8300   alive    1.9.4+ent  2          elwood    934.333µs
e294c42b-efce-4c3b-796b-0de93fc51d1f     ip-192-168-100-203.jake          18.188.194.17:8300   alive    1.9.4+ent  2          jake      1.241941ms
e294c42b-efce-4c3b-796b-0de93fc51d1f     ip-192-168-100-60.jake           3.140.201.212:8300   alive    1.9.4+ent  2          jake      2.867721ms
e294c42b-efce-4c3b-796b-0de93fc51d1f     ip-192-168-100-97.jake           3.135.63.108:8300    alive    1.9.4+ent  2          jake      1.127126ms
```

Figure 4.18 – Members of joined data centers

So, we've covered two different methods of joining the Consul data centers for federation. The WAN Gossip method is available in open source Consul; however, care does need to be taken when implementing this method. If you recall from *Chapter 3, Keep It Safe, Stupid, and Secure Your Cluster!*, WAN Gossip encryption requires only an encryption key that is managed at the Consul server cluster. By linking our data centers using the advanced federation and RPC, we're able to dramatically improve the security posture utilizing mutual **Transport Layer Security** (**TLS**) certificates for authentication of the peer, as well as encryption of the traffic.

Summary

Well, I truly hope you enjoyed that chapter. We've seen how Consul is able to, in many real-world scenarios, scale to incredibly large numbers of nodes. We've seen how *read replicas* and, more importantly, the use of *Autopilot* enables us to scale our server clusters to support networks with tens of thousands of nodes. I believe I've even heard recently of a company running Consul in production with over 100,000 nodes! Centralizing the operation for all services on a system that large can be challenging for sure, so enabling namespaces not only helps scale the application usage within Consul but also the dependency on humans to operate the system as it grows. Finally, we've looked at federating our Consul clusters in a couple of ways to help span our service availability across data centers. As I've said since the beginning, Consul enables communication across components and services. Now those communication paths are no longer bound by scale or geography.

At this point, we should have a decent understanding of the operation of Consul and how the various components within the cluster interoperate. In the next few chapters, we're going to dig deeper into the Consul use cases so you can start applying all of this knowledge to value-generating services! The first use case addresses Consul service discovery, the very foundation of the Consul service offering.

Section 2: Use Cases Deep Dive

Now that we understand how Consul works, let's take a deeper look into how to apply Consul to our primary use cases. We touched on the use cases in *Chapter 1, Consul Overview – Operation and Use Cases*; however, it's time to dig a bit deeper and see them in action!

This part of the book comprises the following chapters:

- *Chapter 5, Little Bo Peep Lost Her Service*
- *Chapter 6, Connect Four or More*
- *Chapter 7, Animate Me*
- *Chapter 8, Where Do We Go Now?*

5
Little Bo Peep Lost Her Service

This is it! The moment we've all been waiting for! All of those horrible dad jokes and puns are about to pay off as we actually start putting all of this tremendous knowledge to use! At the very foundation of every Consul use case is the discovery and sharing of service information. We're ready to focus our energies more on *why use Consul*, instead of *what is Consul*. To understand the why for this initial use case, let's look at the age-old nursery rhyme of *Little Bo Peep who lost her sheep*. However, as you may have guessed, instead of losing sheep, we're losing our service!

Little Bo Peep is tending to her flock of sheep, all free on the hillside. It is easy for her to monitor the sheep while they are confined to their pen. They can only roam so far, and as long as she controls those gates of the pen, there's no problem! As time goes on, she starts to learn that her sheep can provide much more bountiful wool, not only to her but to others, when they are left to roam. Furthermore, as she allows those sheep to nomadically roam, she finds providing food and water for the sheep is dramatically more efficient. So, what does she need to manage these sheep without corraling them into their pen?

- **Location**: Of course, she needs some sort of method to track and locate the sheep. She can solve this by installing **Global Positioning System (GPS)** trackers in each one of them so she can locate them if necessary.

- **Health**: She needs to know that her sheep are healthy! If one isn't feeling well, she needs to tend to it, and just as important, make sure nobody shears that sheep for its wool!

- **Discovery**: So she knows where they are and how they are feeling, but how is she going to let other people find the sheep? Will she just announce it periodically, hold up a sign, or maybe offer a hotline with celebrity guests?

As we dig into this chapter, we're going to see just how Consul solves these three key challenges for Little Bo Peep and her lost sheep, I mean, service.

Technical requirements

Within this section, we'll be deploying a full Consul cluster to **Amazon Web Services (AWS)** utilizing Terraform. If you've already gone through *Chapter 2, Architecture – How Does It Work?*, set up your AWS account, downloaded Terraform, and cloned the repository, you're in great shape!

There are no specific steps throughout this chapter; however, it may be very useful to have a simple running Consul cluster (such as that from *Chapter 2, Architecture – How Does It Work?*), along with the services provided, to navigate and play with the services!

Dude, where's my service?

The first of Little Bo Peep's challenges that we're going to address is simply to determine the location of her sheep, or in this case, her service. In our analogy, Little Bo Peep has installed GPS trackers in each of her sheep. However, the GPS at its core only provides the longitude and latitude of a given point. It alone doesn't give you any information about that point, only that it exists. If she stops at simply installing the trackers, she hasn't really solved the problem, has she? At a very high level, she can see that a GPS tracker exists in a particular location. However, what she doesn't yet know is which sheep exists at that location. She also needs some sort of receiver to maintain the location of all of those GPS coordinates. This comes down to service definition and centralized service registration.

Every service in every network has some sort of definition. Some aspects of the definition could be very high level, such as a name or a purpose. Other aspects of the definition can be incredibly detailed, such as on what address the service can be reached and on what port it is listening. Historically, this information could be maintained in a database, spreadsheet, or sticky note in somebody's cubicle. Regardless of where it was kept, however, it needed to be fully defined. Consul is no different.

Every service that Consul is aware of is defined by, well, a service definition. Now, some might say, *wait, you said this stuff was automatically discovered – why do I need to configure it?* The claim is that these services are automatically **discovered**, not automatically **defined**. The definition comes in the form of a file, either JSON or HCL, that Consul is able to read along with other configuration files. These files can be quite elaborate, or incredibly simple, depending on the service and its function. Let's look at a service definition that has already been employed in our work thus far:

httpd.json

```
{
    "service": {
      "name": "httpd",
      "tags": [
        "python-server"
      ],
      "port": 8080,
      "check": {
        "args": [
          "curl",
          "localhost:8080"
        ],
        "interval": "10s"
      }
    }
}
```

You might notice that this is one file, called httpd.json, with the definition of the httpd service that is deployed in Consul. When the Consul agent starts, it attempts to read every file within the configuration folder. These files could be actual agent configurations, security configurations (such as **Access Control Lists** (**ACLs**)), or service definitions. This affords any organization the ability for the service developers to write, or at least significantly contribute to, their service definition and manage it uniquely. If you don't want an individual file for every service, Consul does provide the ability to combine multiple services into one configuration file. However, care should be taken to not allow those files to become overloaded. After all, can you imagine having 100 services running on a Consul agent with one configuration file?

OK, let's pick apart this file a bit and see what it consists of, which honestly isn't much. We have a name for our service, `httpd`, and a custom tag, `python-server`. We can add more tags if we like, allowing us to identify different service versions, whether an instance is a primary or a backup for a service, or anything else. By default, the service name also serves as the service identifier within Consul. However, in some cases, you may want unique instances of a service on a single node, for example, if we had multiple versions of our `httpd` service and distinguished them through metadata or tags. We can't have the same name pointing to multiple unique service iterations, so you may optionally populate a service ID to address this scenario. We are identifying a TCP port that the service is available on, and a simple health check and check interval to be sure the service is actually operating. More on the health check later.

> **Important Note**
>
> As we progress further, you'll see how we can use the common **Domain Name Service (DNS)** to discover our services. For this reason, it is highly recommended that your service names and any associated tags comply with DNS hostname restrictions.

As you can see, there isn't a great deal of information that is really needed to define a service. That being said, the amount of optional information is quite extensive. All options are available within the official documentation, `https://www.consul.io/docs/discovery/services`, but many of those configuration parameters are for more advanced use cases that we'll get into in the following chapters. In addition to the `name`, `tag`, `port`, and `id` values that we mentioned earlier, I wanted to call out a few additional useful parameters for a basic service definition:

- **Address**: Typically, whatever address is assigned to the node on which the agent resides will also be used as the address to advertise the availability of the service. If the nodes are multi-network, this would be an important parameter to ensure the service is available on the correct network interface.

- **Meta**: Everything seems to be meta these days! This parameter allows you to define up to 64 key/value pairs to better describe and manage your services. Note that each key is limited to 128 characters in length, and each value can be no longer than 512 characters.

- **Token**: If you want to get super granular and pinpoint the control of individual services, the service definition can contain an ACL token for specific permissions. This is also required only if registering the service with a specific Consul namespace.

In a typical Consul system, services are deployed, removed, or updated quite frequently. If you think about the software development life cycle, the gestation period for new services or applications can be as little as a few weeks. In many of these cases, updates to the service definition may be required for new versions, tags, and so on. Ideally, every time a new service or application is released, a corresponding Consul service definition file will accompany the application.

As that new application, or new version of an existing application, is deployed, the service definition would be updated as well. If you're using a simple orchestrator such as Nomad, it can manage the corresponding service definition as part of the application definition and be deployed automatically. However, in traditional systems where you are actively selecting on which nodes the application is to run, whatever process is used to push the application should also be pushing the corresponding service definition.

Once the new definition is available, the associated Consul agent does require a simple reload in order to recognize the adjusted configuration. Much like most configurations, dynamic adjustments to the services aren't picked up automatically. However, reloading the agent is quite simple, via the `consul reload` command, or via an API call:

```
$ curl --request PUT $CONSUL_HTTP_ADDR/v1/agent/reload
```

In a more dynamic world, the Consul API provides a great amount of functionality as part of this process. It can't help you deploy your applications; however, every service definition and check can be provided to the Consul agent utilizing the Consul API. This can be very useful for environments where the service definition, or any aspect of that definition (including health checks), requires modifications post application deployment. Registering the service with the API does require some additional information, however, to ensure that Consul is aware of which node the service is expected to reside on. Note the difference between the following `httpd-api.json` for the Consul API and the original `httpd.json` definition we used previously. The `Datacenter`, `Node`, and `Address` fields are required in order to inform Consul on which node the service is expected, while `CheckId` and `ServiceId` are required in order to link the service and check definitions. This enables the ability to push service and check definitions uniquely:

httpd-api.json

```
{
    "Datacenter": "dc1",
    "Node": "ip-192-168-100-118",
```

```
  "Address": "192.168.100.118",
  "Service": {
    "ID": "httpd2",
    "Service": "httpd2",
    "Tags": [
      "python-server"
    ],
    "Address": "192.168.100.118",
    "Port": 8080
  },
  "Check": {
    "CheckID": "httpd2",
    "Node": "ip-192-168-100-118",
    "ServiceID": "httpd2",
    "Status": "passing",
    "Definition": {
      "tcp": "localhost:8080",
      "interval": "10s"
    }
  }
}
```

Important Note

Although defining and managing services via the API can be incredibly useful for programmatic management of Consul, it must be noted that should the Consul agent restart, any catalog entries performed via the API are not persisted.

External services

But Rob, what if I want Consul to be aware of services that aren't embedded directly into the Consul system? This isn't a problem for Consul; however, it does require some additional configuration options, similar to what we had to do with the API service definition. Let's consider the situation that throughout our life cycle, we want to provide a centralized GitHub Enterprise service to our developers but enable them to utilize Consul for the discovery of the GitHub service. I'm going to use the **GitHub** hosted service platform as an example, but as it's only a URL, this definition can easily adapt to a multitude of scenarios:

github.json External Service Definition

```json
{
    "Node": "github",
    "Address": "www.github.com",
    "NodeMeta": {
        "external-node": "true",
        "external-probe": "true"
    },
    "Service": {
        "ID": "vcs",
        "Service": "vcs",
        "Port": 80
    },
    "Checks": [{
        "Name": "http-check",
        "Definition": {
            "http": "https://www.github.com",
            "interval": "30s"
        }
    }]
}
```

Note that within the service definition, we're also including the node and check information.

The important aspect of this definition, unique from what we've seen before, is the existence of the `external-node` and `external-probe` parameters, both set to `true`. This information informs Consul that the service needs to be monitored by some sort of external service manager. Funnily enough, *external service manager* happens to be the name of a simple binary daemon that can run alongside all of your Consul agents to manage any sort of external query. Before we get to that, though, let's get our service registered.

As not only the service resides outside of Consul, but also the node on which that service lies, Consul requires a registration of the node, service, and any additional checks via the API. We can do this by utilizing **cURL** and the GitHub service definition JSON file from the preceding code:

```
$ curl --request PUT --data @github.json http://<consul_http_
address>:8500/v1/catalog/register
```

With our service registered, Consul is just going to assume that all is well in the **vcs** service world and make it available within the service catalog:

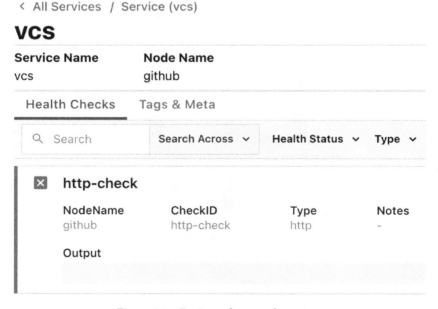

Figure 5.1 – Registered external service

So, we can see that our service has been registered; however, it isn't showing as healthy yet. This is because the Consul agent is looking for an external service monitor to monitor, um, that external service. When the Consul **External Service Manager**, or **ESM**, is running, that too will show up as a service within the Consul catalog. With that service, we can now see that the external GitHub service is now reporting to be in a healthy state:

Services 4 total

🔍 Search	Search Across ∨	Health Status ∨	Service Type ∨

✅ **consul-esm**
1 Instance

✅ **vcs**
1 Instance

Figure 5.2 – Consul External Service Monitor registered

When we look at the service details, we can see not only our HTTP check for GitHub returning healthy but also an **External Node Status** check returning in a happy and healthy state:

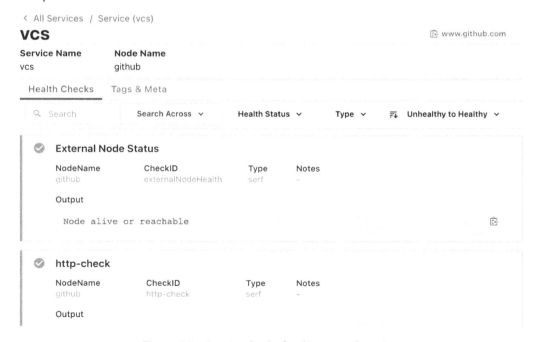

Figure 5.3 – Service checks for the external service

At this point, we've defined an external service within Consul that is not only registered within the service catalog but is also being monitored to ensure the service is in a healthy state prior to providing the catalog information to any consumers.

If you want to download Consul ESM onto your own systems and play with the service, simply log in to one of the Consul client nodes of your choice. At the command shell, execute the following commands:

```
$ curl -O https://releases.hashicorp.com/consul-esm/0.5.0/
consul-esm_0.5.0_linux_amd64.tgz
$ tar -zxvf consul-esm_0.5.0_linux_amd64.tgz
```

This provides you with the consul-esm binary, which can be configured much like a standard Consul agent, using an HCL configuration file! I've provided a very simple configuration file. If you wish to play with ESM, just make sure you modify the local_private_ip parameters and insert the proper parameter with your node:

---

consul-esm.hcl Configuration File

```
log_level = "INFO"
instance_id = ""
consul_service = "consul-esm"
http_addr = "<local_private_ip>:8500"
datacenter = "dc1"
```

Now you have a configuration file, you can run consul-esm and use it:

```
$ sudo ./consul-esm -config-file=./consul-esm.hcl
```

If you're playing along, as you certainly should be, your external service for the VCS should be registered as *healthy* within your Consul cluster. If you would like to check out the Consul ESM source code or other documentation, it is all available open source on GitHub!

```
https://github.com/hashicorp/consul-esm
```

Now that we know how to define both internal and external service. Little Bo Peep is able to determine which sheep are hers, and where they reside. She also needs a way to ensure her sheep are healthy. After all, an unhealthy sheep needs to be tended to! Let's take a look at how we ensure our sheep, I mean services, are healthy.

Major Tom, are you OK?

Whenever we deal with communications, there is almost always some level of feedback that is required to ensure that the receiver is available. This is the entire instigator of the phrases we all know so well, such as *can you hear me now?*. Anybody who has had to deal with network load balancers or system monitoring should be no stranger to the term *health check*. This is exactly what Consul can employ within every service definition. Although a health check isn't a required part of a service definition, I can't imagine why anybody would want to define and deploy a service without one!

In the realm of network load balancers or system monitoring, health checks were defined and orchestrated by some centralized form of intelligence. This means that if a load balancer managed 10 services running across 100 different nodes, the load balancer would need to track 1,000 health checks. This can not only cause some scaling issues as the services grow but it also (at least slightly) increases the amount of network traffic.

Many monitoring systems today utilize remote agents for health checking, addressing both the scaling issue as well as network impact. Consul operates in a similar fashion as the monitoring system, utilizing the Consul agent deployed alongside the service to manage and perform all local health checks. Unlike a centralized monitoring system, however, if there is a failure on a Consul-managed service, or a Consul node/agent, the burden is not solely on the server cluster. Recall that as all of our agents are gossiping, they learn about the service availability of their peers for reporting. Therefore, with Consul, every agent that is managing services also manages and executes the checks that are associated with the health of the respective services. Let's take a look at what checks we can implement for that purpose:

- **Script**: As it sounds, a script check executes a predefined custom script in order to perform some function to validate the health of the service. In our example with the `httpd.json` service definition, we are using a script check utilizing the Linux `curl` utility along with some corresponding arguments (`localhost:8080`). Although I'm using a common utility as the script, in most cases the script is something very custom that performs an in-depth query of the service health. Along with the script, you can specify the interval at which the script is run, as well as any timeout necessary to accommodate for script failure. Note that script checks are not enabled by default, so the configuration of the agent must permit script checks. This can be performed through `enable_script_checks` as well as the `enable_local_script_checks` Boolean parameter. Consul recognizes the health of the service based on the returned value from the script:
  - `0`: Service is healthy.
  - `1`: Service is in warning status.
  - Anything else: Service unavailable

> **Caution**
>
> The `enable_script_checks` field permits scripts defined with any method, including the API. If you haven't properly secured your cluster, this enables remote entities to execute scripts on the Consul agent. It is highly recommended to have any script checks defined and implemented local to the agent and exclusively use `enable_local_script_checks`.

- **HTTP**: As you might be able to guess, an HTTP-based check performs an HTTP request to ascertain the health of the associated service. By default, this is a `GET` request; however, the check does support other HTTP methods as part of the configuration. If specific HTTP headers are necessary, these can also be specified as part of the configuration via the `headers` field. Similar to the script-based check, the configuration can specify both the timeout that Consul will wait for a response to the HTTP request, as well as the interval at which requests are made. Additionally, as we've already seen the importance of securing our communication, HTTP checks do support configuration for TLS in the event that the application utilizes a secure HTTP endpoint. Any `2xx` response from the HTTP check is considered to be a healthy service.

- **TCP**: If you've ever utilized Telnet to check to see whether an application is listening on a port, then the TCP check is for you! The TCP check simply attempts to connect to a host and a TCP port that the application should be listening on. Of course, similar to the HTTP and script-based check, the configuration can specify both the timeout that Consul will wait for a response to the HTTP request, as well as the interval at which requests are made. You may be recognizing a pattern here. If the Consul agent is able to connect to the host and port, the service is considered to be healthy.

- **TTL**: This check is purely based on the **Time-To-Live** (**TTL**) of the last known state of the application. Within the HTTP and TCP health checks, Consul queries the service to validate availability. With the TTL check, however, Consul will listen for HTTP updates to see whether the service is healthy. In most cases, this is utilized only when the service can self-report its status, as either **pass**, **warn**, or **fail**. If the service doesn't report within the TTL value, then Consul will assume the service is no longer available.

- **Docker**: This can be useful to ascertain the health of services or applications that are running within Docker containers without any exposed interfaces on which the TCP, HTTP, or script-based checks can be used. For a Docker-based check, Consul interacts with a Docker host utilizing the Docker HTTP API or a direct socket. Consul directs Docker to spin up a container application that will check the service health, report the health to Consul, and then terminate the container. As part of the configuration, you can specify the interval at which this check is run, but no timeout for this one. Note that similar to the script check, this does require advanced permissions on the Consul agent, therefore the `enable_script_checks` Boolean configurations must be set to `true`.

- **gRPC**: Some applications that support **gRPC remote procedure calls** may also support a gRPC-specific health check endpoint. This is definitely a more specific integration. However, the check supports the health validation of a single service, or multiple services as reported by the gRPC server. The definition of a gRPC-based check has both the `timeout` and `interval` parameters we have seen in other checks. Additionally, similar to the HTTP check, the gRPC health check supports TLS configuration for secure networks.

- **H2ping**: This check provides similar functionality as the HTTP check; however, this supports a very simple ping frame delivery defined via the HTTP2 protocol. Just like the HTTP check, the H2ping check provides the ability to configure the interval at which the checks are executed and the timeout to wait before declaring a check as failed. Unlike the HTTP check, however, H2ping relies upon a TLS connection by default, but it can be disabled by setting `tls_skip_verify` to `true`.

All of the aforementioned checks support optional fields, such as `notes` (to provide human-readable information about the check) and `token` (to provide a specific ACL token for the check to utilize). Additionally, in many instances, we don't want our health checks to just operate as a *one-and-done* operation. It can often be useful to identify how many passing checks need to occur before we trust the service to be available. Additionally, we may want to ensure that a single check failure doesn't necessarily negate the entire service availability. For these scenarios, each check supports a `success_before_passing` field as well as a `failures_before_critical` field. Finally, it should be noted that Consul supports the definition of multiple checks within a single configuration. For the most part, our conversation has revolved around checking for the health of our services. We should also be checking the health of the nodes on which our Consul agents reside. The application of multiple checks is perfectly suited for this purpose as we can then check memory, CPU, disk space, and so on. The following uber-check provides an example of multiple node-level checks (note how there is no service ID associated with the checks):

node-check.json

```
{
  "checks": [
    {
      "name": "mem",
      "args": ["check_mem.sh", "-limit", "256MB"],
      "interval": "5s",
      "timeout": "2s"
    },
    {
      "name": "cpu",
      "args": ["check_cpu.sh", "-percent", "90"],
      "interval": "10s"
    },
    {
      "name": "disk",
      "args": ["check_disk.sh",
        "-path", "/",
        "-threshold", "80"
      ],
      "interval": "10s",
```

```
      "timeout": "5s",
    }


  ]
}
```

This assumes that we have simple scripts created that evaluate the resources, taking into consideration the limits and thresholds as variables. If any one of these scripts results in 1, the node will be considered in a warning state. In order for the node to be deemed healthy, the output of all scripts must be 0.

There we have it. As you can see, Consul supports a myriad of health check protocols to ensure service availability. The heart of Consul's functionality resides in its service catalog; therefore, accurate understanding of the health of the services associated with a particular agent is a critical function of the system.

So, at this point, we have a service catalog, and we're able to validate that our services are healthy and ready to be utilized. Next, we're going to see how our applications can discover those services.

Methods of discovery – no lawyers required

Whenever I hear the term discovery, I can't help but think about all of the lawyer shows and movies that reference the process. Thankfully, we don't require lawyers as part of the discovery process, but there are several methods available to discover what services our catalog offers. Let's see some of them in the following sections.

DNS

The easiest way for any application to discover the location of services is the same way they've been doing it for years, utilizing DNS. By default, Consul offers DNS on port 8600 on every agent within the Consul cluster. Every node and service can be queried utilizing a **Consul** domain. For example, if we want to find the address(es) for the nodes hosting the httpd service, we can execute the following command (using the IP address of a Consul node, of course):

```
$ dig @192.168.100.20 -p 8600 httpd.service.consul
```

Focusing on the answer section of the query, we can validate that two addresses have been returned on which the `httpd` service can be found:

```
;; ANSWER SECTION:
httpd.service.consul.  0    IN    A    3.20.237.142
httpd.service.consul.  0    IN    A    13.59.70.71
```

Hey Rob, everybody knows that DNS utilizes port 53, why don't you just use that for Consul DNS? This is an excellent question and is associated with the privileges under which Consul is running. In order to attach an application or service to port 53, a well-known port, the application requires privileged access. I don't recommend operating Consul (or most applications) with that level of access within any production environment. *Well, how are we going to make use of this now?* Don't worry, young Padawan…there are a few options here.

Within most systems, there exists a plethora of DNS-based services that may be deployed not only to manage the name resolution for that particular node but also to help that node find other nodes within the network. This includes utilities such as BIND, dnsmasq, Unbound, and systemd-resolved. As some of these solutions may result in only resolving the Consul domain, there is a parameter within the Consul agent configuration entitled recursers. This enables Consul to send any DNS queries that it cannot resolve up to another DNS.

API

Compared to the utilization of DNS, accessing services via the API is incredibly simple. However, from the application side, this does require some level of development for service discovery. Along with the API, Consul provides an extensive filtering mechanism in order to manage the criteria utilized to discover the service. Attributes on which we can filter include tags for the service and node, metadata for the service and node, addresses, and the data center. In the following example, we're filtering based on the `ServiceTags` that are associated with our `httpd` service:

```
curl http://18.189.17.157:8500/v1/catalog/service/httpd
'filter=ServiceTags == "python-server"' | jq
```

The result of such a query is quite extensive, but it is all provided in JSON format, which has shown to be quite easy to parse for both man and machine (and maybe cyborg):

Truncated output of Consul API Service Query

```
[
  {
    "ID": "c83108da-f84c-20cf-7882-386e91ee2b27",
    "Node": "ip-192-168-100-168",
    "Address": "13.59.70.71",
    "Datacenter": "dc1",
    "TaggedAddresses": {
      "lan": "13.59.70.71",
      "lan_ipv4": "13.59.70.71",
      "wan": "13.59.70.71",
      "wan_ipv4": "13.59.70.71"
    },
    "NodeMeta": {
      "consul-network-segment": ""
    },
    "ServiceKind": "",
    "ServiceID": "httpd",
    "ServiceName": "httpd",
    "ServiceTags": [
      "python-server"
    ],
    "ServiceAddress": "",
    "ServiceWeights": {
      "Passing": 1,
      "Warning": 1
    },
    "ServiceMeta": {},
    "ServicePort": 8080,
    "ServiceEnableTagOverride": false,
    "ServiceProxy": {
      "MeshGateway": {},
      "Expose": {}
```

```
    },
    "ServiceConnect": {},
    "Namespace": "default",
    "CreateIndex": 24,
    "ModifyIndex": 24
},
```

Both the DNS and API capabilities enable applications and humans to query Consul for service locations. But what if we want to be able to send updates to services out from Consul? This is where Consul **watches** can come in handy.

Consul watches

In addition to these methods, Consul also offers the ability to create custom watches to monitor changes to any number of parameters within the Consul system. Each **watch** is paired with an appropriate **handler** that will execute whenever the watch notices a change. Although the Consul watch can observe multiple different parameters, we're going to focus on the following:

- `Services`: This watches any changes to the service catalog. If any new service is added, or if a service is removed from the catalog, this watch will execute the associated handler. To observe this watch from the Consul command line, execute the following command: `consul watch -type services`.

- `Service`: This watches any changes to a particular service within the service catalog. Watching a particular service provides all of the details associated with that service, including the location(s) on which the service is available and the status of any associated health checks. Any changes to the service status or location will result in the execution of the associated handler. To observe this watch from the Consul command line, execute the following command: `consul watch -type service -service httpd`.

- `Checks`: This is the most useful watch I've found in this scenario and provides the results of the checks. You can query the results of all checks within the system or focus on checks that are associated with a single particular service. To observe this watch from the Consul command line, execute the following command: `consul watch -type checks -service httpd`.

As noted previously, all checks can be operated at the command line, which is a great way to learn more about the operation and results of the check. The results of the check are JSON formatted to facilitate parsing of the data, which can be useful for alerting other components of the service availability. For something such as the Consul `services` watch, the results are very high level, as in the following example:

```
{
    "consul": [],
    "httpd": [
        "python-server"
    ]
}
```

Meanwhile, the output for the service specific watch is quite more elaborate. The following example results have been truncated for brevity.

```
{
    "Node": {
        "ID": "494c2e14-5442-0339-6bbd-fa84254f5aa5",
        "Node": "ip-192-168-100-124",
        "Address": "3.128.24.217",
        "Datacenter": "dc1",
        "TaggedAddresses": {
...
```

If we want to find which nodes are running the `httpd` service we are using in our examples, we can easily parse the JSON result from the query. Additionally, we can add a `passingonly` Boolean parameter to indicate that we only want to retrieve nodes that are hosting healthy instances of the `httpd` service:

```
consul watch -type service -service httpd -passingonly true |
jq ".[] | .Node.Address"
```

Of course, these examples are all interactive with the Consul service, and it only provides the results of the watch. The following example specifies a `checks` type watch that will call the `check-handler.py` Python script whenever the check is executed:

```
watches = [
  {
    type = "checks"
    service = "httpd"
    args = ["/etc/consul/consul.d/check-handler.py"]
  }
]
```

In order to notify the external applications or services of the check results, including the location of our healthy services, the handler is able to post this information out to a multitude of services. One potential option is to update a messaging queue that services are subscribed to, such as **RabbitMQ** or **ActiveMQ**. Once Consul posts the service information, any component that is listening on the respective queue can retrieve the information.

There we have it! Throughout this section, you've learned three different ways to obtain information about the services within the Consul catalog, starting with DNS, which appeared to be the most straightforward, leading to the direct API integration, through to a notification mechanism that can notify just about anything or anybody!

Summary

I hope you enjoyed this chapter...the topic of defining and managing our services isn't a horribly complex one, but it is critically important for the healthy operation of any dynamic network. A huge advantage of our cloudy world isn't the existence of cloud computing; it is how being able to quickly create and remove resources changes our human operational model. Unfortunately, that flexibility does mean our services have the ability to migrate and roam amongst the clouds. Using our basic example, we were able to register our `httpd` service within the Consul system, ensure that the service is healthy, and review the different methods that can be used to discover that service. Now would be a good time to take a break and play with the Consul system, as the next topic, *service mesh*, has quite a few more working pieces to it. Hang on...things are about to get meshy!

6
Connect Four
or More

This is it! All that you have read so far has led you to this point. I know I've said that
before, but it is true. Everything you have read in this book has led you to this point.
Regardless, welcome to the world of service mesh. Any engineer with a background
in routing and networking appreciates the power of the service mesh. For years,
we have focused on connecting machines, allocating or identifying the addresses, and
segmenting them into subnets, switchports, and route tables. This functionality is still
required, of course; however, much of it is now defined as Infrastructure as Code and
is programmatically managed by public or private cloud operators. Now we have
a chance to bring the conversation up a level, quite literally, so we are no longer
connecting strings of numbers, but we are connecting the services directly. In this chapter,
we're going to be meshing with the following topics:

- Connecting our services within the mesh – In the last chapter, we learned how to
  discover our services. Now we are going to learn how to get them communicating
  with each other through the service mesh.

- Connecting our services outside of the mesh – It is unrealistic to think that
  everything you could ever interact with would reside within the service mesh.
  Therefore, we need to look at how to enable and manage connections with
  external entities.

- Securing the mesh – If all the mesh did was allow service-to-service communication, it would be pretty cool, but not very useful in my opinion. Once you have a service mesh enabled, you can uniquely secure, and easily disable and enable, each service connection.

- Mesh to mesh – Much like the assumption that all services will exist in a mesh, it is also unrealistic to assume that there will be a single mesh to rule them all! As we extend our network and services, we will need to connect multiple service meshes, providing a mesh of meshes. Yup, we're going to get meshy!

The progression we've laid out here isn't only good for education, but it provides a logical path for service mesh adoption within any network as well. Without any further delay, or puns, let's connect four, or more!

Technical requirements

In this section, we'll be deploying a full Consul cluster to **Amazon Web Services** (**AWS**) using Terraform. If you've already gone through *Chapter 2, Architecture – How Does It Work?*, set up your AWS account, downloaded Terraform, and cloned the repository, you're in great shape!

Within the ch6 folder of the repository, we have a similar structure as for the ch2 folder, providing the code for creating the image and, of course, creating the Consul cluster. If you would like to follow along, use ch6 to create a new Consul cluster, which includes some of the service configuration files you'll utilize within this chapter.

Before we walk through the code and functionality within this chapter, it might be useful to revisit *Chapter 2, Architecture – How Does It Work?* and review the steps required to deploy and access our infrastructure.

Connect services, not addresses

Whether you take the red pill or the blue pill, the network that we live in can't be refuted. The medium or the base layer connection isn't always recognized. After all, for most of 2020, and at least half of 2021, many of us have been living through connected machines. Nearly every phone out there today resides on some sort of computer network, identified by an assigned address. However, as consumers, we don't really care about that address, at least not directly. The functionality that provides and maintains that address only exists to allow our applications to communicate. Looking at the communication path as a whole, there are other aspects beyond that assigned address that are required, for example, some sort of physical connection and identification. Each of these aspects functions as a layer in what is called the **Open Systems Interconnect** (**OSI**) model.

Unless you've been playing with cables and switches for the last several years, this may be a new concept. So, before we talk much more about the service mesh, I want to get a baseline of networking laid out. Don't worry, we aren't prepping for a Cisco Certified Internet Expert exam.

Understanding the network connection

When our applications communicate with each other, the messages that are transmitted need to travel through a variety of mediums. When we open up our favorite video conferencing software, the application connects with a server on the other side. I think we all can understand that. But what facilitates that communication? We can say it just goes over Wi-Fi, but what does that really mean? Well, the application software and the Wi-Fi are just layers within the communication path. Thankfully, we aren't dealing with the number of layers of an onion, or an ogre. And we're even going to simplify the number of layers for our discussion as well.

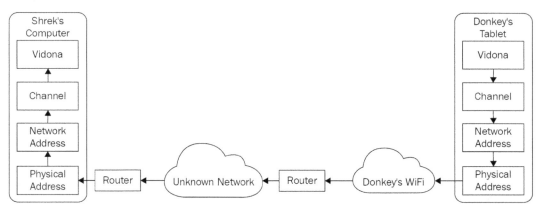

Figure 6.1 – Vidona connection over Swampnet

Take a look at *Figure 6.1*. Due to a pandemic in the swamp, Shrek and Donkey haven't been able to see each other for quite some time. However, they've discovered a new application, **Vidona**, that allows them to *see* each other virtually. When Donkey calls Shrek, the message originates in **Vidona**, the application in use. That message flows through a channel and is wrapped up in an envelope with Shrek's address, and Donkey's return address. The message then flows down to the physical layer. Although we can't *feel* Wi-Fi signals (wouldn't that be wild if we could?), the Wi-Fi radio exists physically. In order to send the message on the Wi-Fi radio, it is packaged in another envelope, with a different address representing Donkey's physical tablet.

Now we have two addresses: a **network** address and a **physical** address. The **physical** address doesn't change (usually). It is associated with Donkey's tablet, and it helps the tablet talk to the components it is *physically* connected to (even if wirelessly). Meanwhile, the **network** address identifies Donkey's tablet as a globally unique entity, which is what Shrek's computer will use to respond to Donkey's tablet. Note a very important distinction here; Donkey's **Vidona** application requires knowledge of Shrek's network address in order to connect, and the network address changes over time. Furthermore, if Donkey and Shrek want to keep their communication secret from the eyes of Lord Farquaad, **Vidona** must develop its own security mechanisms. This may be challenging for the developers of **Vidona** alone, and when other applications need to participate in this secure communication, it becomes even more difficult. Ultimately, however, the goal is to provide a secure connection at the Application layer, the **Vidona** layer, as shown in *Figure 6.2*:

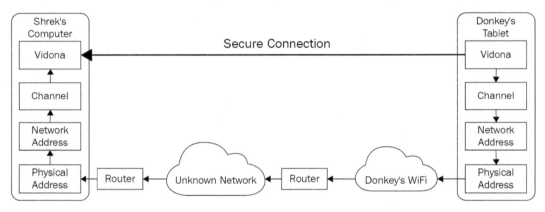

Figure 6.2 – Vidona secure connection

The important distinction between *Figure 6.1* and *6.2* is that although the underlying network flow does not change, the **Vidona** application is able to communicate among peers in a seemingly direct and secure manner. As **Vidona** controls both sides of this particular example, utilizing its own built-in security mechanisms may be sufficient. However, once you mix applications and services, this continuity breaks down.

As part of our conversation moving forward, we're going to be referring to various layers within the networking world. At the beginning of this chapter, I referred to the OSI model, which consists of seven layers, as shown in *Figure 6.3*:

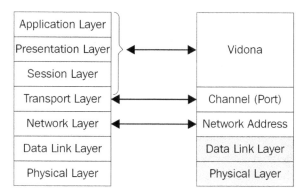

Figure 6.3 – Network OSI model

For our purposes, we are going to focus primarily on the top three layers, Application, Presentation, and Session, which we will collectively refer to as the **Application layer**. The Transport layer, as well as the Network layer do come into play as well, however, these are at the machine level of communication.

With that basic understanding of the networking stack, let's take a look at why simply relying on that stack just isn't enough for today's cloudy world.

Why traditional networking isn't enough

In the realm of zero-trust, we can't just implicitly assume that anything operating on a single machine is friendly. Nearly every week we hear about a security event that originated with the assumption that an entity was trusted, and yet it was compromised. The concept of zero-trust, or least privilege security, focuses on the identity and the security of what we truly care about, the application. If our server, or system, is hosting multiple applications, we need a standardized way to uniquely identify, and secure, each of those applications. To illustrate this point, I'm going to revisit the common phone number analogy.

When I was a young lad, I would on occasion participate in activities that could be considered questionable at times. Luckily, we weren't all carrying computers with GPS trackers in our pockets; however, my parents had the phone numbers for the houses of all of my friends. If I was going to Mike's house, my parents would know how to reach me by way of the phone number associated with Mike's house. However, when they called, there was no control over who would answer. Among our friends, we were able to have some fun with this level of ambiguity. For example, when our friend Gina would call Mike's house, I might answer the phone. Now we were all very close, but even now sometimes it can be difficult to identify somebody by their voice over the telephone. It was a lot more challenging then, with lower quality audio on landlines! So, here I was having an active conversation with Gina; however, she believed she was speaking with Mike. We can continue this practice today; however, it is more difficult to pick up somebody's mobile phone than it was to just lift the handset in their house. In this analogy, Gina and I are the applications, and Mike's house is a host. Gina was trying to reach Mike by way of his house phone number, but simply by being there, I was able to impersonate Mike. The phone number may be trusted and clearly identified with the host, but that doesn't ensure the identity or the security of the application.

This story demonstrates how simply assuming a **layer 4 address** as the identity of any application imposes risk. In a private data center, when the house was securely locked with guards placed at the door, identification by address wasn't that unreasonable. However, in the environment of the public cloud, the existence of legitimate and illegitimate address masking, and especially in any sort of migratory workload environment (such as Kubernetes), the address alone just isn't good enough. So, let's look at how we can solve this problem using tunnels.

Tunneling to safety

The concept of network tunnels and **Virtual Private Networks** (**VPNs**) is nothing new. However, in every one of those instances, the connection is based on the layer four address of the components involved. We can connect our application machines using a series of point-to-point VPNs, but that isn't going to alleviate the need to identify and secure the **application**. In reality, the use of VPNs actually increases the need for application identity instead of relying on simply the address. However, we can still look at the VPN as a model for where we want to be.

When you establish a VPN in most cases, there is a software application that connects to some remote site or service. When that connection is made, your device now has a new IP address, valid on the destination network, and all machine traffic flows through that connection. The beauty of VPNs is that the information traversing the VPN is encrypted and secured, from the VPN client to the VPN server. This was, and continues to be, a great way to protect your traffic from prying eyes. The use of VPNs historically has been primarily for enterprise networks. Lately, many people today utilize VPNs to not only encrypt their traffic, but also to hide their layer four address identity. We've already determined that for true secure operations, we need to validate the identity of the application, so the use of a VPN may actually hinder, rather than help, that effort.

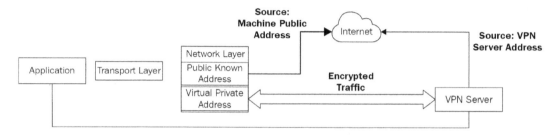

Figure 6.4 – VPN traffic pattern

Now, I'm not saying that VPNs don't have their place in this world. When you need to connect **machines** in a secure manner, they provide a vital function. However, we aren't talking about connecting machines; we are connecting our **applications**. That being said, the concept of the VPN extends nicely if we can bring that connection up a layer to the application. This is exactly what the Consul Service Mesh provides via Consul Connect.

The Consul Connect functionality **connects** identity-verified applications and encrypts the traffic between those applications. We've already seen in *Chapter 3, Keep It Safe, Stupid, and Secure Your Cluster!* how the Consul server cluster verifies the identity of the agents and encrypts the RPC traffic utilizing **mutual TLS (mTLS)**. Consul Connect allows us to extend that same functionality to every Consul managed service. Let's take a few minutes and revisit Consul and mTLS.

Remember the days of old, when we were having fun with certificates in *Chapter 3, Keep It Safe, Stupid, and Secure Your Cluster!*. As part of that process, we were utilizing the Consul server as the **certificate authority** (**CA**) to sign and manage the certificates (it's an authority, what else would it do?). In an effort to reduce complexity, we were using common certificates for each of the three servers. Additionally, all client agents were using a common certificate. At the time it was mentioned that in a production environment, each Consul agent should have its own unique certificate. When one agent receives a public certificate from another, that certificate provides the **identity** of the sender. We know we can **trust** that identity, because the certificate is signed by some authoritative source. These certificates not only provide the identity, but also the keys necessary for the encryption and decryption of the messages between the agents. If this method of mTLS is secure enough to validate Consul agents and encrypt the traffic, why not use it for the services that Consul manages?

With Consul Connect, each service is assigned a unique certificate. As you can see, the use of certificates is quite important when working with Consul. Not only do we have certificates for the agents to enable secure agent communication, but we also have certificates for the services. All of these certificates can be quite challenging to manage, but the benefit of true application identity and encryption far outweighs that challenge. Luckily, much like in *Chapter 3, Keep It Safe, Stupid, and Secure Your Cluster!*, we are able to utilize Consul as the CA for our applications. Consul does support Vault as the CA as well; however, for our purposes, we are going to focus on utilizing Consul. Regardless of who the CA is, Consul can orchestrate the rotation and distribution of the certificates, which can dramatically simplify the operational aspect of Consul. Unlike *Chapter 3, Keep It Safe, Stupid, and Secure Your Cluster!*, however, we want our applications to utilize the certificates for identification and encryption. Unfortunately, many applications have not been developed to accommodate special certificates and encryption. There are development libraries for custom application integration; however, proxy servers are wonderfully suited for this purpose and require no additional development.

By default, Consul supports two proxy servers for this purpose. For education and development, Consul provides its own proxy server. This proxy server operates at layer 4 (remember our model), often referred to as the **Transport** layer. Utilizing the built-in proxy, Consul creates tunnels based on the TCP port that is being used by the application. This doesn't bring us all the way to the application, but it's a great improvement over simply the IP address of the machine. Within a production environment, however, the Envoy proxy provides additional functionality, including the ability to manage traffic at the HTTP level (for HTTP-based services). This brings us directly to the actual application and enables unique identities for multiple services simply based on the URL being used to call the service. Within a production environment, this layer 7 functionality also enables more advanced topics such as service routing or service splitting. We won't be covering these topics here, but at a high level, they provide the ability to customize how traffic is divided between multiple instances of services. This functionality is critical for blue-green, or canary deployments, which are useful for testing out new versions of the application before spinning the wheel of chance and upgrading all of the nodes.

Regardless of which proxy server is utilized, we can think of them as tunnel entry and exit points within the mesh. Additionally, because the proxy is often co-resident with the actual application, we can completely disconnect our service from the outside world requiring all communication to flow through the proxy server. For example, in a typical environment with a web server, that server would be listening on some externally facing address and port. If you recall the `httpd` service that we utilized in the previous chapter, this service was listening on all interfaces simply due to the nature of the service itself. Within Terraform, we start that service as part of the `remote-exec` provisioner identified by the following command:

```
nohup python3 -m http.server 8080 &
```

We didn't define what interface (address) that service would be listening on, so it listened on all addresses available on the system. To prove this, we can run a simple `netstat` command:

```
$ netstat -anp |grep 8080
```

The first line returned should be similar to the following:

```
tcp        0      0 0.0.0.0:8080            0.0.0.0:*
LISTEN      1304/python3
```

This shows us that our Python-based web server is listening on port 8080, with the address 0.0.0.0. This value represents all available addresses and interfaces. With that configuration, any device that can reach this machine can access port 8080. To increase security, we're going to change how we run that httpd server for this exercise:

```
nohup python3 -m http.server 8080 --bind 127.0.0.1 &
```

With the additional bind configuration flag, we are telling the Python http server to **only** listen on 127.0.0.1, which is the localhost of the machine. To validate this, we can run our handy dandy netstat command again, providing us with the following result:

```
tcp          0        0 127.0.0.1:8080              0.0.0.0:*
LISTEN          7954/python3
```

With the server listening only on the localhost address, no outside application can access the server. Unless, of course, they start getting meshy. So, now I have to ask, are you ready to get meshy with me?

Within the repository folder for *Chapter 6, Connect Four or More*, I've placed a new httpd.json file, as well as a curl.json file. These are going to represent our new meshy services, and as part of the Terraform build, these files will be placed on the Consul clients.

The httpd.json service definition hasn't changed much from our original definition, with one main exception. We now have a connect configuration block, which identifies that this service is going to utilize a sidecar for connectivity. That sidecar is the proxy server.

httpd.json

```json
{
    "service": {
      "name": "httpd",
      "id":"httpd-1",
      "tags": [
        "python-server"
      ],
      "port": 8080,
      "connect": {
        "sidecar_service": {}
      },
```

```
        "check": {
          "args": [
            "curl",
            "localhost:8080"
          ],
          "interval": "10s"
        }
      }
    }
```

The `curl.json` file is a bit different, as the `curl` utility doesn't really provide a service, it is more of a client application. However, `curl` is a tremendous command-line client for any web server!

curl.json

```
{
    "service": {
      "name": "curl",
      "tags": [
        "curl-client"
      ],
      "connect": {
        "sidecar_service": {
          "proxy": {
              "upstreams": [{
                  "destination_name": "httpd",
                  "local_bind_port": 3579
                }]
            }
        }
      }
    }
}
```

Notice that the `connect` portion of this configuration is slightly more involved. For the sidecar service, we need some additional parameters to tell the proxy server what service `curl` will be connecting to (`httpd`), as well as the local port on which the proxy server should listen for requests. Within my system, and I encourage you to play along, I have the `httpd` service running on one Consul client, with the `curl` process on another. However, when I try to look at these services within the Consul UI, I find they are not in a healthy state!

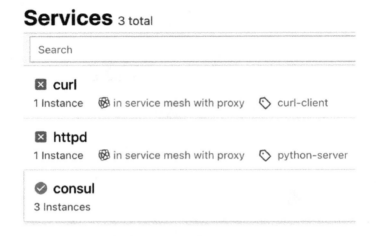

Figure 6.5 – Unhealthy httpd and curl services

What we can see in *Figure 6.5*, however, is that the services are **in service mesh with proxy**. Well, this is new! I wonder if the fact that the service is in a service mesh has something to do with the lack of availability. If we look at the service instances, we can see that there is this thing called **Connect Sidecar Listening**, which we also haven't seen before. This looks to be the problem, and it obviously appears similar to the sidecar stuff we've been playing with.

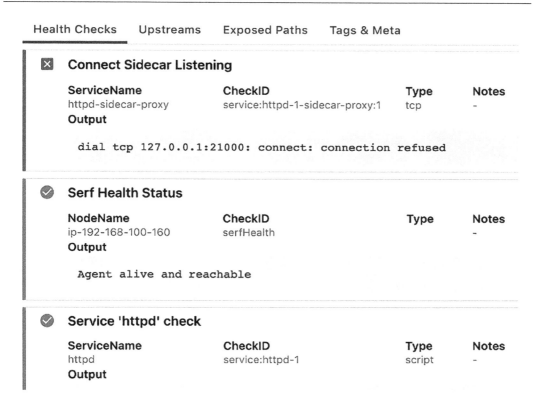

Figure 6.6 – Unhealthy sidecar

In order to get this service healthy, the sidecar needs some attention. What we've done so far is register an httpd service in the catalog, and we've told Consul that the service will be accessible via proxy. However, we haven't done anything to start that proxy! When utilizing the service mesh, every application (unless it supports the mesh natively) requires its own proxy service. Therefore, if Consul isn't aware of any running proxy service that is paired with the service, the service will not be in a healthy state. To fire up the proxy server for the httpd service, all we need is a single command:

```
$ consul connect proxy -sidecar-for httpd-1 > ~/httpd-proxy.log
&
```

What we've done here is tell Consul to issue a connect request, utilizing the internal proxy, operating as a sidecar for the httpd-1 service. *But wait, I thought it was just httpd!* How astute you are . . . I'm very impressed! When creating the proxy services, we utilize the **ID** of the service, not the **name** of the service. Any output of that request is being redirected to a log file, and we're running the command in the background. Once we run that command, the service is now healthy and happy within the Consul system.

Now, we have to get the `curl` service up and running. I find it strange calling `curl` a service since it's only a simple command line, but for our purposes, we are going to imagine `curl` is a client application accessing some web server. To get the proxy server for the `curl` service running, we're going to run a similar command:

```
$ consul connect proxy -sidecar-for curl > ~/curl-proxy.log &
```

Now we have proxy servers for both `curl` and our web server running. From the Consul UI, both services are healthy, and we can even see the topology for the service connections. The topology isn't very exciting at the moment (like curling), but as the number of services grows, having a view of what services are connected is critical to understanding the communication paths. The path that we've established is represented in *Figure 6.7*:

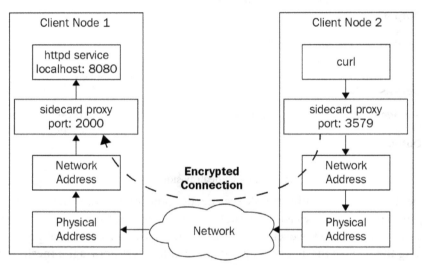

Figure 6.7 – Connected services via proxies

Let's walk through this flow a bit from the client's perspective. We have a proxy server running alongside a service we are calling `curl`. The definition of that service, specifically the `local_bind_port`, identifies a port (`3579`) that the proxy should be listening to for messages. Any messages that the proxy receives on that port are going to be sent upstream to the `httpd` service. Note that we have no port or address for that service configured; all we have is the service name used as the destination. On the server side, we have another proxy server running alongside our `httpd` service. We haven't defined a port for that service, the proxy server dynamically allocates the port when it starts. For this instance, the proxy server allocated port `21000`. Because the `httpd` service is running on port `8080` on the **internal** localhost interface, the proxy server knows to forward requests to that port.

To demonstrate that we have the mesh working properly, we're going to pretend that *we* are that client application. So, emulating a client request, from the client machine, we are going to reach out to port 3579. That message will flow through both proxy servers to the httpd service:

```
$ curl http://localhost:3579
```

The response is a directory listing of the server node from /home/ubuntu, where the Python web server is running. If you want to double-check, you can use a utility called tcpdump:

```
$ sudo tcpdump -nnvvXSs 1514 -i any port 21000
```

If you run that command on the server node that is running the httpd service and then issue the curl command, you'll see packets received and sent. Don't worry, you aren't expected to be able to read the packet information. As a matter of fact, you won't be able to! The proxy servers are utilizing certificates provided by Consul to encrypt that communication. Although there is no encryption configured on the web server or the curl utility, the communication between the two devices is successfully encrypted. Furthermore, the routing that we did was all based on the service name, without any address configurations! At this point, you deserve a congratulatory virtual hug.

It is important to note at this point that the Consul server cluster isn't participating in the actual exchange of information between the client and the server. The data between these applications flows through the proxies directly to each other. This path of data is often called the **data plane**. Very inventive, I know. However, the information about those proxy servers that is necessary to establish that data connection is managed by Consul. All of the activities that Consul performs in this effort take place on the **control plane**. After all, these are the communications necessary to **control** your cluster! With that centralized control point, we have the ability to create rules to either permit or deny communications between services. Within the world of Consul, we call these **intentions**.

State your intentions

Now that we understand how our applications will communicate over the service mesh, we need to understand how to configure and manage that communication. We may want to think that all of our services will live in harmony, and only those that should speak to each other shall do so. However, as altruistic as this idea is, it isn't very realistic. Therefore, we need a mechanism to easily define what applications can, and more importantly cannot, access one another within the Consul system. For this, we use what is called a Consul **intention**.

The nice thing about where we are with Consul services is that each service is defined by a human-understandable name. I would strongly recommend clear and distinct service names within any system simply to avoid ambiguity. However, that effort provides some additional comfort in that we can now specify intentions via the service name. If you've ever had to deal with any sort of IP-based firewall, you should find this very reassuring. Similar to an IP-based firewall, however, the Consul intention defines the connection based on the source of the communication method, and the destination service or application. The granularity of that control, however, depends on what protocol the service is defined to use.

The service definition enables the ability for the administrator to define the protocol on which the application resides. There are four options for the protocol: `tcp`, `http`, `http2`, and `grpc`. However, for intentions, we are going to focus on the first two, `tcp` and `http`.

The `tcp` protocol definition provides greater flexibility than the `http` protocol, as it extends supportability beyond `http` servers. If you want to utilize Consul for a database, for example, this uses the `tcp` protocol. In this instance, the identity of the application encoded within the mTLS certificate is a combination of IP address and TCP port. Intentions that manage these connections are called layer 4 intentions. If we look at the Application layer, layer 7, we can set intentions based on the actual URL utilized within the communication. This is necessary for service routing and splitting, which we discussed previously regarding the proxy servers.

One additional important distinction between layer 4 intentions and layer 7 intentions is when intention updates become effective. Without delving too deep, when there is a TCP connection between client and server, that connection is established and remains connected for the life of the communication path. For HTTP, however, the connection is only active for the request and response for the information. Therefore, layer 7 intentions may become effective much more quickly than layer 4 intentions. Consul doesn't tear down existing connections, so only when a new connection request is made will the intention have an impact.

Regardless of the layer at which the intentions are enforced, the configuration procedure is identical, and it only requires three pieces of information: the name of the service, the source of the communication request, and whether to allow or deny the connection. Before we get into the definition of the intentions, I want to revisit the *ACL* section of *Chapter 4, Data Center (Not Trade) Federation*.

If you recall, there are two methods in which Consul will run as a default for ACLs. We can either implicitly deny all requests and only specify the access that is permitted, or we can implicitly allow all requests and specify which connections we want to deny. Following along with the zero-trust model, critical to any security posturing, the implicit deny should be in place for any production environment. Consul intentions follow the same model, where we can either implicitly accept all network connections or implicitly deny all network connections. Whatever policy you have configured for your ACLs, the same policy applies to your intentions. For example, if your ACL policy is *deny all*, then by default any allowed service connections need to be explicitly defined. As we've already shown traffic flowing through the mesh, we can see that we don't have a *deny all* ACL policy in place.

So, in order to play with intentions, we'll be using that same `curl` to `http` server connection that is already established. Utilizing the CLI, let's take a look at the services that are registered within the Consul catalog:

```
$ consul catalog services
consul
curl
curl-sidecar-proxy
httpd
httpd-sidecar-proxy
```

Note that we actually have services listed for each of the sidecar proxy servers. However, our **intention** is to manage the actual services, `curl` and `httpd`. If we want to prevent communication between `curl` and `httpd` using the command line, it is quite simple:

```
$ consul intention create -deny curl httpd
```

The CLI responds, acknowledging the creation of the intention:

```
Created: curl => httpd (deny)
```

With this in place, let's go ahead and try issuing that `curl` command again:

```
$ curl http://localhost:3579
curl: (52) Empty reply from server
```

As you see, we are no longer able to connect: `Empty reply from server`. I get the same reaction when I ask if a menu item is vegan…empty reply from server. If you observe the logs from the `httpd` server when this happens, we get a very interesting message:

```
[ERROR] proxy.connect: authz call denied: service=httpd
reason="Matched L4 intention: default/curl => default/httpd
(ID: a605dfa9-cdcc-b304-400d-3fe708b603c1, Precedence: 9,
Action: DENY)"
```

From the result, we can plainly see that an attempt was made from `default/curl` to `default/httpd` and it was denied. The *default* value that we see here relates to the namespace that the services, and the intention, applies to. This allows unique intentions to be configured and assigned for individual namespaces within the Consul system. Furthermore, we already know that with the proper ACL tokens, we can ensure that only those that are authorized can create intentions in their own namespaces. We also have a precedence value returned. With any rules in life, there is going to be an order of precedence. There may also be wildcards in place to generalize some of those rules and help with the management. For example, we may want to make sure that certain services on our system *only* receive connections, and never transmit. For example, our `httpd` service should never be initiating connections to any other service. For this purpose, we can wildcard an intention that instructs Consul to deny any request originating from the `httpd` service:

```
$ consul intention create -deny httpd '*'
```

If you aren't implicitly denying all connections within your production environment, this type of rule can be incredibly useful to manage any potential breach of an application. For example, if our `httpd` service was compromised and used to send information to any other service within our mesh, this intention would explicitly prevent such an occurrence from happening. It should be noted as well that wildcard intentions are very low in the precedence list, as shown in *Figure 6.8*. The more precise the intention, the higher the precedence that intention will have in the evaluation.

Source Namespace	Source Name	Destination Namespace	Destination Name	Precedence
Exact	Exact	Exact	Exact	9
Exact	*	Exact	Exact	8
*	*	Exact	Exact	7
Exact	Exact	Exact	*	6
Exact	*	Exact	*	5
*	*	Exact	*	4
Exact	Exact	*	*	3
Exact	*	*	*	2
*	*	*	*	1

Figure 6.8 – Consul intentions order of precedence

Within our Consul system, we have now created two intentions: one to prevent `curl` from communicating with the `httpd` service, and another to prevent any outward communication from that `httpd` service to any other service. Within the Consul UI, all of the configured intentions are visible, and can even be sorted by order of precedence! As the number of services, and therefore intentions, grows, this view can be incredibly useful.

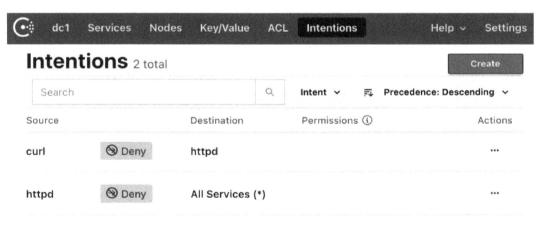

Figure 6.9 – Intentions in the UI

Managing intentions within the UI is quite simple either by creating a new intention or editing an existing intention. Every service within the Consul mesh is provided as a source and destination, and we can either choose to allow, deny, or create an **Application Aware** rule:

Edit Intention

Source   Destination

Source Service   **Destination Service**

| curl ▼ | | httpd ▼ |

Should this source connect to the destination?

→ **Allow**	⊗ **Deny**	≋ **Application Aware**
○ The source service will be allowed to connect to the destination.	◉ The source service will not be allowed to connect to the destination.	○ The source service may or may not connect to the destination service via unique permissions based on L7 criteria: path, header, or method.

Description (Optional)

| Description (Optional) |

| Save | Cancel | | Delete |

Figure 6.10 – Editing intentions within the UI

The **Application Aware** rules add complexity through the use of explicit permissions. Recall that layer 7 intentions provide additional functionality to define intentions based on the path of the URL or even the headers in the request.

Up until now, every scenario that we've been discussing has been based on the event that every application and service exists within the service mesh. What can we do about services that reside outside of the mesh? Well, I'm glad you asked, as that question provides an excellent gateway to the next topic.

Gateways to domination

Wow, such wonderful and entertaining information on the service mesh! But I have a problem: not all of my services are eligible for the mesh. To quote the Finnish-American architect Eero Saarinen, *have you tried gateway?*

To think that every service, and every application, will reside within a single service mesh is quite unrealistic for any enterprise deployment. For this reason, Consul offers the concept of a **gateway.** Nearly every gateway in existence, be they spiritual, philosophical, or physical, connects two realms of existence. Looking at Consul, we can view each cluster as an individual realm, within which services are discovered and can safely communicate with each other. However, we need to consider the following scenarios as well:

- **Multiple data centers** – We know we can connect multiple Consul clusters within different data centers. However, when we leave the mesh, we lose the security of identity and encryption.

- **External services** – Not every service that we need will exist within the service mesh. As we've seen, the mesh requires a proxy server co-resident with the relevant application. Many services, such as hosted databases, don't offer an embedded mesh compliant proxy server (at least, not yet).

- **External clients and applications** – Especially if we are offering services to end consumers, the mesh needs to accept traffic from external entities. There must be a way for external entities to enter the mesh to access the services they are entitled to.

Each of these three scenarios requires a Consul agent dedicated to the function of gateway operation. Additionally, the Envoy proxy must be paired with that agent to provide that tunnel-like connection. As we saw in the previous section, Consul utilizes the Envoy proxy server to manage the connections among services within the mesh. This functionality wouldn't be possible without the tight coupling between Consul and Envoy, a coupling that is required for the mesh gateways. Therefore, Envoy at this point is the only proxy server supported for this purpose. Let's take a look at how Consul applies various gateways in these scenarios.

Mesh gateways

We've seen how we can connect services securely using the service mesh. I like to think about mesh gateways as kind of a *super-mesh*. Utilizing WAN federation, which we played with back in *Chapter 4, Data Center (Not Trade) Federation,* we are able to connect Consul clusters across multiple data centers. The Consul mesh gateway takes that connection and adds not only the encryption to that connection, but also the ability to control which services are permitted to traverse those data centers.

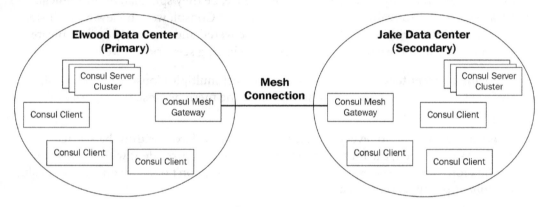

Figure 6.11 – Mesh gateways connecting data centers

Whenever we connect a service within the Consul service mesh, the Consul server cluster manages the certificates for identity and encryption through the **connect** functionality. Connecting clusters through the mesh is no different; however, we all know what happens when we have more than one boss (let alone six bosses Bob, six bosses). Therefore, whenever we are connecting data centers via a mesh gateway, a primary data center must be identified to manage and distribute the certificates. These certificates create a new mTLS session between mesh gateway peers used for identity. It is important to note here that the encrypted traffic that flows through the mesh gateway is still service-to-service encrypted. In other words, the mesh gateway entities do not decrypt the message between services; they only retain the encryption that was already performed at the service endpoints.

Well, what if we don't want all of our services traversing through the gateway? No problem! When the services are connected through the mesh gateway, the Consul clusters recognize those services, and we can write intentions! This gives the Consul administrators the immense power of not only controlling the service mesh within a single cluster, but the entire world! Furthermore, as part of the service definition, we can tell Consul that the **upstream service** (the service we want to initiate the connection to) is available via the mesh gateway. Let's take a look at a modified `curl` service as an example.

Curl.json across the mesh gateway

```
{
    "service": {
      "name": "curl",
      "tags": [
        "curl-client"
      ],
      "connect": {
        "sidecar_service": {
        "proxy": {
            "upstreams": [{
                "destination_name": "httpd",
                "datacenter": "secondary",
                "local_bind_port": 3579
            }]
          }
        }
      }
    }
}
```

Within this slightly altered service definition, we are telling Consul that the `httpd` service that we want to reach is available in the data center named *secondary*. This is a great example of connecting to another Consul data center for access to a service, but what if the intended service isn't within the mesh?

Terminating gateways

With the proliferation of **anything as a service** (*aaS), cloud providers are working very hard to differentiate themselves and remain competitive. At least, that's my theory, which I believe leads to hosted services that would normally require dedicated compute. Even if it isn't true, it sounds logical, and besides, I'm pretty sure I read it on the internet somewhere. Hosted database services are a wonderful example of this functionality, and unfortunately, we can't deploy a Consul agent on a hosted AWS database service. For this purpose, we have the role of a terminating gateway.

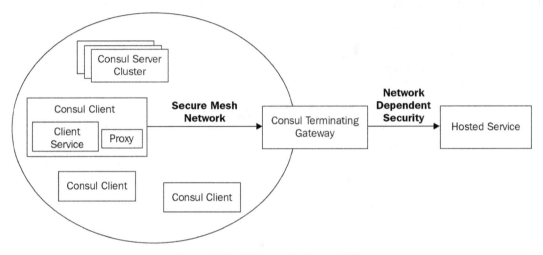

Figure 6.12 – The terminating gateway

Luckily, this gateway isn't wandering through the clouds looking for Sarah Connor. However, what it does terminate is the security and control of the Consul services. All is not lost, however, because if the destination service supports mTLS, the terminating gateway will maintain that secure connection. If the service only supports one-way TLS, the gateway can work with that too; however, this can open the terminating gateway to potential attacks. As the gateway is at the edge of the mesh, unlike the mesh gateway, it does have the ability to decrypt the traffic it receives. Therefore, it is always recommended to tightly control access to the node operating as the gateway. There is a problem though . . . how does Consul know about the services that aren't even inside the Consul cluster?

Recall that as we were discussing service discovery in *Chapter 5, Little Bo Peep Lost Her Service*, Consul is able to contain services within the registry that are not being managed by Consul. These are **external** services, which kind of makes sense, as they are **external** to Consul. Any services that are to be accessible via the terminating gateway must be registered in Consul; after all, if they aren't, how can we discover them? When we configure a terminating gateway, we need to list the services that are available through that gateway. This is the only way that Consul can manage the discovery and connection from within the mesh to those external services. We can still use ACLs and security up to the gateway node; however, once the traffic leaves the gateway, we need to resort to old-fashioned networking, the way our grandmothers did.

Ingress gateways

Next, we have to think about all of those entities that are outside in the great big scary world that want to access the wonderful services that we offer! This requires a method to open the service mesh and invite others to join us. However, we just don't want to open the door and let anything into our mesh. We've worked very hard and have put up with a plethora of horrible jokes to secure our beautiful mesh. The last thing we want is to put the mesh at risk by allowing nefarious traffic through the door.

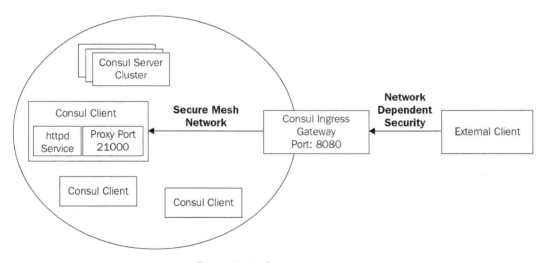

Figure 6.13 – Ingress gateways

Luckily, any service that we would want to be accessible via the ingress gateway should already be registered within the Consul service catalog. However, we don't want to expose ourselves completely as part of the process. Consul allows us to define a listener as part of the configuration for the ingress gateway. This listener tells the Consul agent (and Envoy proxy as well) which port to listen on, and what the associated service is. With this configuration, the ingress gateway can receive traffic on the specified port and forward that traffic to the appropriate destination service.

Now that we can see how Consul can connect internal and external services to the mesh network, let's take a quick look at an environment that exists as individual networks that yearn to be connected.

Meshing with Kubernetes

Throughout this entire chapter, I've only mentioned Kubernetes once, and that was in passing. This is partly due to my loathness of Kubernetes and the number of hours I've struggled with the technology. However, service orchestrators such as Kubernetes (or Nomad) definitely have a place in the world of service mesh. In fact, many people don't think about service mesh unless it is within the realm of one of these application orchestration tools. We aren't going to get into the details of Kubernetes for a number of reasons (some of which I've already mentioned). However, I do want to address why service mesh is so important within this arena.

Application orchestration provides an entire operating environment in which our applications reside. Once deployed within the environment, the primary job of the orchestrator is to ensure that the desired number of instances are operating and that the resources are being managed. The application can reside anywhere within the cluster of machines, so long as any constraints are satisfied. If the application fails, it is up to the orchestrator to restart the instance somewhere within the cluster. Additionally, when performing upgrades of machines, the orchestrators are able to move the applications throughout the cluster to best suit the desired goal. Essentially, the orchestrator, whether it's Kubernetes or Nomad, has full control over the location, and therefore the addressing, of the application. This migratory and ephemeral nature of the applications dramatically complicates the problems of IP addressing discussed earlier.

Within Kubernetes, we have our own server cluster to manage all of the **nodes** (machines) and applications being orchestrated. Today, we have the benefit of the cloud providers offering their own managed Kubernetes services, dramatically simplifying the (often severe) burden of managing the cluster. The cluster can have any number of **nodes**, which provide the compute resources for the applications to be deployed. Each application being deployed and managed is packaged as a **container**. This brings us to the Pod, which is a set of one or more containers that live, operate, and die together. The Kubernetes Pod often shares various attributes among the containers within the Pod. These attributes include things such as storage, and . . . wait for it . . . networking!

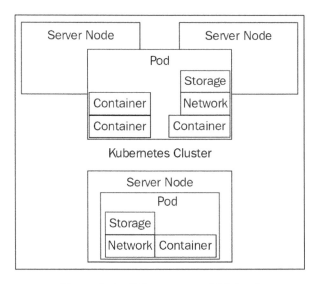

Figure 6.14 – Kubernetes at a high level

When deploying applications to Kubernetes, we use a command called `helm`, which is accompanied by a helm chart that defines how the application is to be deployed. Consul provides its own helm chart for this purpose and is available publicly on GitHub (`https://github.com/hashicorp/consul-k8s`). Utilizing this helm chart, Consul can be installed alongside the relevant Pod(s) within the Kubernetes cluster, facilitating a secure network connection to the Pod. Much like the mesh example outside of Kubernetes, within Kubernetes the Consul Connect feature utilizes the Envoy proxy server as the connection point to the application. The primary difference is that instead of the Consul agent running as an individual process, it runs as a **sidecar** process within the cluster configured with Kubernetes **annotations**. This process is the same regardless of whether the application is the destination for the mesh connection or the source for the mesh connection with an upstream service.

Again, when the topic of service mesh arises, it usually relates to the concept of orchestrators such as Kubernetes. However, I hope I've been able to show you how the applicability extends far beyond the Kubernetes cluster. Kubernetes is an advanced topic in itself and, in my opinion, can be very confusing, and it is exactly that level of complexity that requires technologies such as the Consul service mesh.

Summary

Congratulations! If you've been able to follow this chapter and deal with my horrible jokes, you have a foundational view of the Consul service mesh. The topic of service mesh is a very lengthy one, and I truly hope that this has provided a decent start to that meshy journey! We discussed (well, I wrote, and you read) a bit about traditional networking and why the service mesh is such a valuable tool in today's multi-cloud world. We were able to see a bit of how the mesh worked, connecting our services securely, and how to ensure some services did *not* connect. We had a brief overview of how to use the mesh principles to connect data centers, and how to support services outside of the mesh. Finally, we had a very high-level view of how Consul operates within the world of Kubernetes.

If the topic of service mesh really gets you tingly, I strongly recommend you continue your education. HashiCorp has several tutorials on the topic, and as the need for the mesh has grown, countless blogs and videos have been produced. The topic is far too broad for a single chapter, but I do hope that I was able to provide some clarity on the subject. Next, we're going to look at how Consul can help us minimize, or eliminate, those dastardly tickets for network changes through automation!

7
Animate Me

We've already talked at great length about the Consul cluster's awareness of the services and applications deployed within the system, along with their corresponding health. Why should Consul keep all of this information to itself? A big part of the successful existence of any network, human or machine, is the sharing of information. Throughout this chapter, we're going to look at a few ways that Consul is able to share that information with third-party components in order to improve our lives through automation:

- Key/value (KV) store – Consul utilizes the distributed KV store to share information among Consul nodes, but this can also be used by humans and machines to share data and directives throughout the cluster.

- Watches and handlers – In *Chapter 5*, *Little Bo Peep Lost Her Service*, we saw how Consul watches and handlers are able to announce specific changes to the service status. In this chapter, we'll see how this can be used to distribute information to third-party components.

- Consul and Terraform – Although this isn't a book on Terraform, we have been using it to create and manage our infrastructure. The final section of this chapter is going to show how Consul works with Terraform to automate nearly any infrastructure components based on Consul's application and service awareness.

So, whether the title of this chapter conjures up fears of the singularity overtaking humans, or the sounds of an incredible song from the band Rush, please join me as we journey into the world of a more intelligent and self-reliant network.

The following topics will be covered in this chapter:

- The need for automation
- Distributed configuration with the KV store
- Watching your handlers
- Floating with Consul-Terraform-Sync

Technical requirements

Within this section, we'll be deploying a full Consul cluster to **Amazon Web Services** (**AWS**) utilizing Terraform. If you've already gone through *Chapter 2, Architecture – How Does It Work?*, set up your AWS account, downloaded Terraform, and cloned the repository, you're in great shape!

Within the `ch7` folder of the repository, we have a similar structure as for `ch 2 folder`, providing the code for creating the image, and of course, creating the Consul cluster. If you would like to follow along, use `ch7` to create a new Consul cluster, which includes some of the service configuration files you'll utilize within this chapter.

The need for automation

Within the world of information technology, we live by tickets. Thankfully not speeding tickets or violation of other laws, but methods in which work can be requested and addressed. Entire product lines and organizations have been built around these processes, and rightfully so. I have a colleague who raves about a book called *The Phoenix Project*. If you haven't read or listened to it, I would strongly recommend it. The entire fictional story focuses on the process problems with IT and how technology might be all well and good, but the *process* of making changes often dictates the success or failure of any project.

As we look at our networking and security equipment, the importance of following this ticketing system increases by orders of magnitude. It's not like ordering the wrong virtual machine size or the wrong operating system. The slightest misstep can disrupt entire organizations and cause millions of dollars in damage and lost revenue. Therefore, any changes to this equipment need to be officially requested, evaluated, tested, and tracked throughout the change management process. You don't have to work in this field to have felt the challenge of DNS settings taking down entire networks and preventing our kids from attending virtual classes!

This level of fragility, for lack of a better term, also means that when these changes to the network or security realms are required, we move very methodically. This pace often results in the frustration of project managers, along with executives who have made commitments to fictitious timelines, all across the globe. However, when we move fast and make mistakes, that doesn't make them happy either. The only solution here is the automation of these changes and removing the responsibility from these error-prone and inferior humans! We're going to see how Consul helps us get there, but first I want to address the elephant in the room (a gold star for those that just remembered the psychedelic scene within Dumbo).

Running with scissors

When discussing automation with friends and colleagues, my sanity has been questioned. Yes, there are several seeds that drive that questioning, but on automation alone, I fully admit that running with scissors can be dangerous and scary, especially if there is no sheath on the scissors. As humans, however, we have evolved to discover ways to maintain a level of safety without completely sacrificing performance. Helmets and seat belts are great examples, and within our information technology world, I would include ticketing systems and change control processes.

Embracing automation within our networks doesn't require us to forget all about change control, but it does require us to think differently about it. Through automation, our changes can be more prescriptive and consistent, especially when defined as code. However, that doesn't remove the need for historical tracking or the ability to revert changes. Furthermore, I find it unrealistic to think that all changes should be automated all the time. For example, adding or removing a new node in a load balancing pool or adjusting the log level for all instances of a service in a data center sound like perfect candidates for automation. Exposing every newly deployed service to the public internet may carry a bit more risk. Therefore, careful consideration should be given when determining what should and should not be automated within the system.

Distributed configuration with the key value store

As we've read, one of the fundamental protocols of the Consul system is the gossip protocol, enabling each Consul agent within the system to share information with all others. The information is centrally located and managed by the Consul server cluster, but all agents (who are properly authorized if required) have access to that data. Through a very simple KV interface, Consul provides this ability to distribute intelligence to all entities interacting with the system. Although this provides a very easy way to distribute information, Consul isn't a fully featured distributed data store. Objects within the KV system have a maximum size of 512 KB, and the data only persists to the local data center. However, this doesn't prevent us from using it to do some pretty cool stuff.

To see the benefit of applying the Consul KV system for configuration adjustments, consider the situation of troubleshooting a distributed application. After all, nearly everything we deploy these days consists of a set of services, as opposed to a giant monolithic application. When problems occur, which they always do, we'll find that the log data that we need is only available at elevated levels. From my horrid recollection of doing this in the past, the process usually goes something like this:

1. Acquire the IP addresses for each machine/component that needs to be adjusted.
2. Open a bunch of terminal windows on my machine.
3. SSH into the first machine, usually using some shared PEM key that everybody on the team has.
4. Edit the configuration file to change the log level.
5. Reload the service and check the logs to make sure that the service is running again with the new log level.
6. Log out of the machine and proceed to the next one in the list.

Once the debugging session is over, we have to go back into the machines and revert the changes we've made. Configuration management tools such as Ansible, Puppet, or Chef can help with these changes as well, but those changes often require updates to extensive scripts, which brings in additional risk.

Interacting with the Consul KV system is quite simple. Much like everything else we've done with Consul, we can utilize the command line, the API, or the web interface. The path to the key and value is represented as, well, a path, not unlike the folder and file structure that we all know and love. This not only lends itself well to the organization of the data but also facilitates logical path based API calls, which are all REST-based. We're going to walk through these actions, and if you would like to follow along, all of the code for creating the image and Consul system is located within the GitHub repository we've been using. Of course, there is a new folder for this chapter at `https://github.com/PacktPublishing/Consul-Up-and-Running/tree/main/ch7`.

We can use a previous image that we've built, however, this new build includes some additional functionality that we'll be using in later sections of this chapter. Therefore, I would recommend you create an updated image (`$github_root/ch7/Image-Creation`) as we've done in previous chapters.

With our new Consul system built and ready to go, let's create a couple of values in our KV store. We're going to be playing with environment variables in order to ensure that we are hitting different Consul nodes, so please follow along closely. As an example, I'll be referring to the following Terraform output:

Terraform output

```
Outputs:

CONSUL_HTTP_ADDR = "http://3.128.180.154:8500"
CTS_IP = "3.141.25.114"
Consul_Client_IPs = [
  "18.117.222.65",
  "3.17.161.19",
  "18.222.124.50",
  "18.118.47.132",
]
Consul_Server_IPs = [
  "18.222.216.65",
  "3.128.180.154",
  "18.117.157.22",
]
```

We want to focus on the clients that we've created, as we don't expect our applications to be writing directly to the Consul server nodes! So, let's pick the first client in the list and export that value as the Consul HTTP address:

```
$ export CONSUL_HTTP_ADDR=http://18.117.222.65:8500
```

Note, I'm using the output in the example. You should be using the addresses in your own output. If you want to confirm connectivity, you can execute a consul members command and validate the output.

Now let's write our first value to the KV store:

```
$ consul kv put son/seventh/gift clairvoyance
```

Note that we didn't have to create the path prior to writing our value. As we created our entry, we informed Consul where we wanted the value to be written.

If we want to read this value from the graphical user interface, we can follow the same path under the **Key/Value** section:

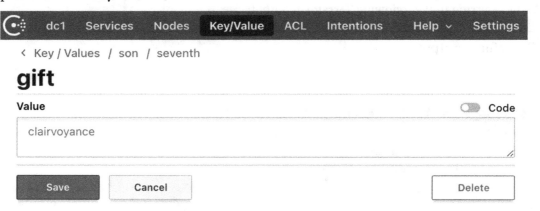

Figure 7.1 – UI access to KV store

Seeing this value within the user interface should give us some reassurance that the value is now available within the cluster, but let's check it anyway using the API. Using cURL, we're going to make an API request to a different client within our cluster:

```
$ curl http://18.222.124.50:8500/v1/kv/son/seventh/gift | jq
```

The output here shows us a bit more information than what is directly visible within the UI:

```
[
  {
```

```
    "LockIndex": 0,
    "Key": "son/seventh/gift",
    "Flags": 0,
    "Value": "Y2xhaXJ2b3lhbmNl",
    "CreateIndex": 194,
    "ModifyIndex": 194
  }
]
```

But wait, that value isn't what we entered! You are correct, this is the base64 encoded value that we've seen in other areas of Consul. This allows more complex data within the values of our KV store to be accessed programmatically. If we want multiple values to be returned within a path, we can set a `recurse` flag on the request:

```
$ curl http://18.222.124.50:8500/v1/kv/son/\?recurse=TRUE   | jq
```

Let's take a look at some of the other values that are provided as part of this output:

- `LockIndex` – Using Consul Sessions, we are able to lock access to particular parameters within the KV store to prevent inadvertent overwriting or conflicts.

- `Key` – This is the full path to the key from the root of the KV store.

- `Flags` – Defaulting to `0`, this parameter provides the ability to assign an unsigned integer to the value entry. This value is opaque to Consul but may help provide additional information (such as version) to an extensive KV implementation.

- `Value` – This is the value of the key encoded to base64.

- `CreateIndex` – Here we have a numerical value that provides some insight into when the value was created in relation to other keys or updates.

- `ModifyIndex` – This is a good value to watch as part of the output, as an intelligent state machine can determine whether the value has changed since the last time it was queried.

This information is always returned via the API, however, if you would like to see this detail within the command-line output, we can use the `detailed` flag:

```
$ consul kv get -detailed son/seventh/gift
CreateIndex       194
Flags             0
Key               son/seventh/gift
LockIndex         0
```

ModifyIndex	327
Session	-
Value	clairvoyance2

Note that I updated the value, and the `ModifyIndex` parameter recognized this change.

We should note here that writing passwords and sensitive information to the storage might not be the best idea in the world. For secrets management, we really should be utilizing a world-class secrets manager such as HashiCorp Vault. That being said, there may be information written into this storage that requires some level of access control. Recall that through our ACL system, we are able to secure paths and functions within Consul. By combining our ACL system with the KV store, we can create a distributed storage system for each application and ensure that only that application (with the proper ACL token) can access the data store.

This is wonderful! Using the KV store within Consul, we can write all sorts of information to the storage, and make it available to specific consumers of the data or applications. *But Rob, my applications have no idea about the Consul API or command line, and you're out of your mind if you think we're going to adjust our development path for that.* Well, you are correct, I am out of my mind, but that has nothing to do with Consul. Allow me to introduce you to Consul Template.

Consul Template

The application of Consul Template may be one of my favorite things about Consul. However, it isn't just restricted to Consul, as Vault implementations can utilize the same functionality. With Consul Template, configuration files can be populated by Consul directly utilizing parameters as if it was, well, a template. I guess you probably saw that coming.

The functionality of Consul Template isn't embedded within Consul itself, however, so it must be downloaded and executed as its own service. The binary (`consul-template`) is available at the HashiCorp releases website if you would like to download it yourself, however, when we built new images for this chapter, installation of the Consul Template binary was part of the process (`https://releases.hashicorp.com/consul-template/`).

Also, as part of this build, I've created a very simple template file and placed it in the /tmp directory. This template file is a very simple text file and utilizes the information that we've populated within our KV store that identifies the gift of the Seventh Son of a Seventh Son:

SeventhSon.tpl

```
Seventh Son of a Seventh Son is a story about a young boy born
with the gift of {{ key "/son/seventh/gift" }}. It has been
popularized in literary form by Orson Scott Card, and musically
by Iron Maiden through their epic album of the same title.
```

Within the template, we can see where Consul is directed to insert the value for the son/seventh/gift key.

In order to utilize the template and create a file, consul-template can be run manually. However, we're talking about automating as much as we can through Consul's intelligence. Why would we now resort to manual operations? So, instead of manually running consul-template, we kicked off the application as a background operation as part of the remote-exec operation performed when we created the infrastructure:

```
nohup sudo consul-template -consul-addr=\"${self.private_ip}\"
 -template \"/tmp/SeventhSon.tpl:/etc/consul/seventh_son.txt\"",
```

This command instantiates the consul-template application, identifying the IP address to use to connect to the Consul service. For our infrastructure, the consul-template function is running on the same machine as Consul, but it can live elsewhere and still make use of Consul's data. We are telling consul-template to load the template file tmp/SeventhSon.tpl and create a new text file (/etc/consul/seventh_son.txt) with that template. As the application is running in the background, consul-template will monitor the value for the identified key, and when that value is changed, consul-template will update the target file as well. We can try this in our own little environment.

If you've been playing along, you've already set a value of clairvoyance for the son/seventh/gift key. If consul-template is working as designed, we should already have our file created. To validate SSH to one of the clients that are available, execute the following command:

```
$ cat /etc/consul/seventh_son.txt
```

We should see the text displayed as specified in the template, with the value clairvoyance inserted:

```
Seventh Son of a Seventh Son is a story about a young boy born
with the gift of clairvoyance. It has been popularized in
literary form by Orson Scott Card, and musically by Iron Maiden
through their epic album of the same title.
```

Let's take a look at what happens to that file when we change the value. Update the value of the son/seventh/gift key. Feel free to do this through any of the methods we practiced previously. Once you've updated the value, check the seventh_son.txt file again to see if your updated value has been reflected. Now, we can imagine that this simple text file is a configuration file for an application. Our templates can be as comprehensive as necessary to address a multitude of configuration files, all managed dynamically and automatically updated based on the values within our Consul KV store!

If your neurons are firing with excitement, please keep it rolling! There is so much more that can be done with Consul Template beyond the scope of what we've covered here. The full documentation, and source code for the project, is also available on GitHub at https://github.com/hashicorp/consul-template.

As we've now covered the usage of KV storage, along with the use of Consul Template, let's move on and take a look at utilizing Consul watches and handlers as another method of automating the configuration of third-party components.

Watching your handlers

Way back in *Chapter 5, Little Bo Peep Lost Her Service*, where Little Bo Peep lost her service, we touched on the functionality of Consul watches and handlers. Recall that a Consul **watch** observes any number of parameters within Consul, and based on value changes, can execute a **handler** to perform some function. Within *Chapter 5, Little Bo Peep Lost Her Service*, we focused on being able to watch the status of a service, in order to inform a third party about service health. We can use this same functionality, but instead of watching service status, we can watch the status of a parameter within the KV store.

For example, if we wanted to watch our key, using the direct command line, we would execute the following:

```
$ consul watch -type=key -key=son/seventh/gift
{
    "Key": "son/seventh/gift",
    "CreateIndex": 194,
```

```
"ModifyIndex": 237,
"LockIndex": 0,
"Flags": 0,
"Value": "Y2xhaXJ2b3lhbmNl",
"Session": ""
}
```

We can see that the output of this command provides us with the nearly identical output from the API when we were reading the value for our key. However, we aren't talking about manual operation; we want…say it with me…AUTOMATION!

Within the Consul agent configuration, we can define the watches that we want to be running on the platform along with a handler. The handler performs some function based on the results of the watch, and can either call an HTTP interface or execute some custom script. Therefore, as the values for a watched key change, Consul will execute that handler to proactively notify some device. For example, the following watch section added to the agent configuration would result in the execution of an update_configs.sh script:

Sample watches configuration

```
watches = [
    {
        type = "key"
        key  = "son/seventh/gift"
        handler_type = script
        args ["/etc/consul/updates_configs.sh"]
    }
]
```

This particular watch pays special attention to a particular key, however, recall that we can define a watch with keyprefix, enabling us to watch an entire path within the Consul KV store. This provides us with the ability to create multiple keys for a particular configuration, and when any one of those keys change, the watch will execute the handler, and perform the desired operation.

As I hope you can see, the power possible within Consul watches and handlers in order to adjust component configuration dynamically is quite immense. We've now seen multiple ways to utilize the Consul KV store to update configuration. Applications can receive updated configuration values through a `pull` method using the API. We can automatically adjust configuration files utilizing the `consul-template` function. And of course, we can push KV information into nearly any system that can be controlled via scripts or an HTTP interface. Next, we're going to look at how we can expand beyond KV and utilize Consul to dynamically adjust our infrastructure components using Terraform!

Floating with consul-terraform-sync

Throughout this book, we've been looking at how Consul can help us discover and connect our services, focusing on the higher-level connectivity (remember that model, working at layer 4 or above). Although this is exactly what we need for our service-to-service communication, the fact that the entire networking world relies on IP addresses to discover, route, and protect our systems has not gone away. Firewalls and load balancers remain in place, largely operating on addresses, managing traffic entering or leaving our managed networks. This is often referred to as **North/South** traffic. Historically, these components would rely upon countless health checks to validate service availability or massive configuration files where a simple typo could bring an entire network down. For these reasons, changes to load balancing and firewall components by the hands of an error-prone human require careful scrutiny. If only there was some way to automatically discover and implement updates to these components in an automated and trackable fashion?

We are already familiar with an incredible tool for managing infrastructure that we've been using as we've built and managed our own infrastructure for this book, Terraform. By using the Terraform OSS binary, we've been creating and destroying the Consul systems necessary for our education. If you aren't familiar with Terraform, it can do so much more. At the time of writing, Terraform offers supported integration with over 1,100 infrastructure providers. This includes the big cloud vendors as well as monitoring components, load balancers, firewalls, content distribution systems, domain name systems, and so on. For automation, many organizations integrate Terraform with a pipeline process, however, we're going to look at how we can automate Terraform infrastructure adjustments utilizing the plethora of data provided by Consul.

> **Important note**
>
> Up until this point in the chapter, most of our discussion has been around creating and sharing configuration parameters throughout our services. The Consul integration with Terraform drastically extends that capability and therefore does warrant additional scrutiny as part of the change control process. Although our work here focuses on the open source versions of Consul and Terraform, an enterprise integration enables the employment of fine-grained policies to specify what automation can change, what it cannot, and what requires an additional level of approval. Furthermore, Terraform Enterprise records state changes throughout the process, providing a history of any automated or approved changes.

Much like the **consul-template**, HashiCorp offers an additional standalone binary application called **Consul-Terraform-Sync**. For the purpose of simplicity, I'll be referring to Consul-Terraform-Sync as CTS moving forward. Utilizing CTS, Consul monitors for changes in the service catalog, and based on those changes, issues a request to Terraform to adjust the infrastructure accordingly:

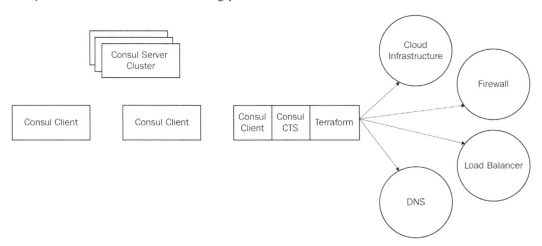

Figure 7.2 – CTS managed infrastructure

As part of the current deployment, we've created a machine specifically for CTS, with the IP address provided in `terraform output` by the variable `CTS_IP`. If you can't find the output from your Terraform run, no problem. Running `terraform output` will provide you with all of the necessary data again. So, let's grab that IP address and SSH into the CTS machine:

```
$ ssh -i <username>-consul-key.pem ubuntu@<CTS_IP>
```

When we built the infrastructure for this chapter, a new directory was created that holds our CTS configuration file. Let's take a look at that configuration file:

cts-config.hcl

```
log_level = "INFO"
port = 8558
syslog {}
buffer_period {
  enabled = true
  min = "5s"
  max = "20s"
}
consul {
  address = "192.168.100.65:8500"
}
driver "terraform" {
  log = false
  persist_log = false
  working_dir = ""
  backend "consul" {
    gzip = true
  }
}
task {
  name        = "find-httpd-nodes"
  description = "Example task with two services"
  source      = "findkim/print/cts"
  version     = "0.1.0"
  services    = ["httpd"]
}
service {
  name = "httpd"
  tag = "python-server"
  # namespace = "service-namespace"
  # datacenter = "dc1"
  description = "Filter for httpd services that have the python-
```

```
server tag"
}
```

We have a few sections within this configuration, so let's take them one at a time.

The first eight lines describe general parameters for the CTS application. The `log_level`, `port`, and `syslog` sections should be pretty self-explanatory, but what is with this `buffer` stuff?

One of the challenges we find with automating changes such as these is that when there is any sort of instability, that instability can have a ripple effect. As humans, we can *sometimes* take time to react to a situation, and by the time we are able to react, the situation has changed. Of course, I say sometimes because this isn't always the case, and we all know that when we don't take time, problems occur. Machines don't usually have this problem of delay. Actually, our problem with machines and software is typically just the opposite, where the applications respond so quickly, we get flapping back and forth. Our buffer configuration helps manage that flapping and ensures that there is some period of time waiting for the system to obtain a steady state before making an adjustment.

The next section identifies the Consul related configuration parameters in the `consul` configuration block; essentially telling CTS where it can find the Consul service. Here we only have the address, however, any certificates or tokens would be called out in this area. As we've already seen how granular we can get with ACLs and tokens, we would want to make sure that the CTS service is utilizing a token that would give it read access to the service catalog.

We know that CTS requires a connection to Terraform in order to initiate infrastructure changes. Optionally, the `terraform` configuration block can specify a pre-downloaded Terraform binary and `path` to be used. However, without this specification, CTS will download a Terraform binary (adhering to the `version` specification if included) and save that binary within the CTS working directory. Logging the Terraform actions, as well as the Terraform backend configuration, should take special attention.

During processing, Terraform can log the interaction with the infrastructure components being managed. This log output may contain sensitive information, and if stored locally on the machine(s) running CTS, could pose a security concern. Therefore, unless you're debugging, it is recommended to not log the Terraform interaction. This same sensitive information is also recorded in Terraform state, which isn't optional. Terraform utilizes that state file to understand the current status of the infrastructure being managed, and evaluate what changes are necessary.

So, what are we to do with this ominous state file? Well, in the first part of this chapter we talked about the functionality available with the distributed Consul KV store. So long as we can keep that state file to a manageable size, less than 512 KB, we can keep that state file within the Consul KV store. This is actually what we are doing with the `backend` configuration sub-block within the `terraform` block. This not only gives us a distributed storage location that we can protect using ACLs but also an indication of changes to the state file (remember the `ModifyIndex` value in the KV response). This does require us to keep the state relatively simple, but we really should be doing that anyway, focusing the Terraform script used for CTS only on the specific infrastructure we need to manage. For example, if I have 10 virtual machines, a Kubernetes cluster, and a load balancer, I should only include the infrastructure code for the load balancer as part of CTS.

All of this talk about state and the Terraform configuration may seem excessive, but the details here cannot be undervalued. To summarize, keeping our Terraform code focused only on what needs to change as a result of the service catalog enables the following:

- Centralized and protected storage utilizing the Consul KV store – we have Consul available already, so why not use it for this purpose?

- Sensitive information in the state file is kept to a minimum, only related to the infrastructure being managed by CTS.

- Faster changes initiated through CTS. With large state files, as a result of large infrastructure definitions, it will take Terraform longer to process the changes.

The next block we want to address is the `task` block. Now we'll finally get to define what we actually want CTS to do! The `task` block identifies the Terraform module that is to be executed as a result of changes to the service catalog. For our purpose, the module is incredibly simple, printing out the service nodes and relevant information. However, a Terraform module can do just about anything within the realm of infrastructure management! A Terraform module can be likened to a function within an application, built for one specific purpose. Some modules can be quite large (such as creating an entire production HashiCorp Vault system), but we've already seen that we want to keep the Terraform operation focused. Therefore, we want to use very specific modules, such as populating the client nodes in a load balancer. The Terraform registry, `https://registry.terraform.io/`, offers a plethora of example modules to be perused.

Our task block simply provides a `name` and `description` for the task, the Terraform module `source` to be used, the `version` of that module, and the `services` to watch that would initiate the call to the module.

If we wish to narrow down the services that would kick off the task, we can use the `service` block. Using this block, we can single out services with particular tags, names, or those that reside in a particular data center and/or namespace.

Note

Note that although we have both task and service blocks within the single configuration file, it is not necessary. Much like the Consul agent configurations, the CTS configuration can be defined within multiple files, all following HCL format, within the same configuration directory.

Now that we've reviewed the configuration, let's take a look at CTS in action! If you haven't deployed the infrastructure, both the image and the Consul cluster, defined within the `ch7` folder, please do so if you would like to follow along.

All of our operations will be performed on the CTS node that was deployed for this chapter's exercises. Using the SSH key that was created as part of the infrastructure build, SSH to the CTS node:

```
$ ssh -i rjackson-consul-key.pem ubuntu@18.191.237.27
```

Our configuration file has been created and placed in `/etc/consul/cts` for your viewing pleasure. In a production system, we would be running CTS as a service on the machine, pointing to a configuration folder for the CTS, task, and service-based configuration. We could have done that; however, all of the work would have been done, and what would we learn that way anyway?!

From the home directory where you've logged in, first list the contents using `ls`. You'll only see a few ZIP files, but this will be your working directory for CTS. Now, let's take a look, or rather **inspect**, what CTS will do if we run it:

```
$ consul-terraform-sync -config-file /etc/consul/cts/
cts-config.hcl -inspect
```

As this is the first time we're running CTS, a few operations will happen in the background. First, CTS will prepare the Consul backend for the Terraform state files. We can see this within our Consul system KV store as well:

Figure 7.3 – CTS within the KV store

Within the folder for CTS, we can see, as shown in *Figure 7.4*, how we have a state file specific to the find-httpd-nodes task that is configured. Note that each task, and therefore module, will create its own state file, and will store those files here within Consul:

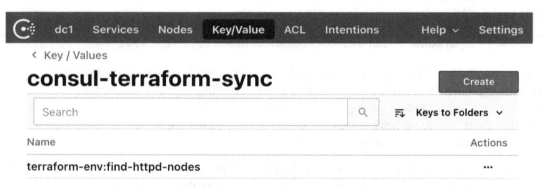

Figure 7.4 – Task state file storage

If we look in our directory, we can also see that we now have our Terraform binary downloaded by CTS, but we also have this folder called sync-tasks. Here is where the magic happens.

```
─── consul-template_0.25.2_linux_amd64.zip
─── consul-terraform-sync_0.1.2_linux_amd64.zip
─── consul_1.9.0_linux_amd64.zip
─── sync-tasks
    └── find-httpd-nodes
        ├── main.tf
        ├── providers.tfvars
        ├── terraform.tfvars
        ├── terraform.tfvars.tmpl
        └── variables.tf
─── terraform
```

Figure 7.5 – CTS content created

Within the sync-tasks folder, we will have a folder for every task that CTS has been commanded with. For us, we just have find-httpd-nodes, the name of our task. Within that folder, we have a bunch of Terraform code! More importantly, we have a terraform.tfvars.tmpl file. If you observe that file, we'll find that it looks a lot like a Consul Template file. If you continue your path with HashiCorp products, get used to these template files. They are incredibly useful, and they pop up everywhere!

When CTS runs, that template file is used to create a new terraform.tfvars file, the file that actually holds the variables used for the Terraform code. All we've done so far is run an inspect operation; basically, a task with CTS to see **what** it would do, but not actually do it. These variables, created by CTS, are fed into the Terraform code that is then used to adjust the infrastructure accordingly.

If you recall, as we've been building our infrastructure for the book, we've run `terraform plan` before creating any infrastructure. The output of `plan` tells us what Terraform intends on doing. If we look at our CTS `inspect` output, we see something very similar:

```
An execution plan has been generated and is shown below.
Resource actions are indicated with the following symbols:
  + create

Terraform will perform the following actions:

  # module.find-httpd-nodes.local_file.consul_service["httpd"] will be created
  + resource "local_file" "consul_service" {
      + content              = <<~EOT
            3.143.4.121
            18.116.20.247
            3.15.226.4
            3.12.146.63
        EOT
      + directory_permission = "0777"
      + file_permission      = "0777"
      + filename             = "httpd.txt"
      + id                   = (known after apply)
    }

Plan: 1 to add, 0 to change, 0 to destroy.
```

Figure 7.6 – terraform plan output

The output shows us exactly what Terraform will do, creating a local file called `httpd.txt`, with the IP addresses of the nodes offering the **httpd** service. We can see as part of the output that we only have one resource that will be added…the file created by the module.

Alright, so now we know what Terraform will do, why don't we give it a shot? Execute the following command, which tells CTS to only perform the Terraform function once. This is always good practice as it will actually exercise the Terraform application of the code:

```
$ consul-terraform-sync -config-file /etc/consul/cts/cts-
config.hcl -once
```

This process doesn't take very long. After all, Terraform is only creating a text file with the outputs. However, if we had a very complex set of modules to be adjusted, this is where we would see the delay. Although we have our buffer settings, we don't want to get into a situation where once the Terraform code has completed, it needs to run again for another change. Proper monitoring of the logging from CTS output would tell us if this were happening.

Now that we've run the command, we can see that our `httpd.txt` file has been created. The file is created in the directory of the task, and we can see the result by issuing the following command:

```
$ cat /home/ubuntu/sync-tasks/find-httpd-nodes/httpd.txt
```

Now we have all four nodes that are running our `httpd` service, provided in a text file. That alone isn't very exciting until we see that it's the file that makes this special; it's the fact that it was created using Terraform! Now pretty much anything that can be performed within a Terraform module can be performed automatically, using service catalog information provided by Consul! When we run the service as a daemon (not a demon), CTS will continually monitor for changes to `httpd` in the service catalog and utilize Terraform to update the infrastructure accordingly.

Summary

Let's take a look at what we've done here. The Consul system has the potential to hold a wealth of information about the services within your network. That information is power! Of course, that's the power to connect securely, but I hope that through this chapter, you've seen how Consul's power can extend far beyond the figurative walls of its gossiping corner. We've seen how the KV system can be used to store and distribute configuration information. We've learned how to construct Consul watches, enabling the notification service changes to other components for configuration management. Lastly, we've seen how Consul interoperates with Terraform, using the Consul-Terraform-Sync capability, to virtually adjust any infrastructure based upon Consul's service knowledge. Now it's time to sit back with a beverage of choice, and dream of all of the things that Consul can do for you!

8
Where Do We Go Now?

I wanted to leave you with one last chapter that will hopefully launch you further into the world of service discovery, service mesh, and automation utilizing the Consul system. This book was merely the start, and I hope it was something that you found both entertaining and informative.

Consul Associate Certification

If you would like to continue to the next level, I recommend the **HashiCorp Certification Programs**, specifically the Consul Associate Exam: `https://www.hashicorp.com/certification/consul-associate`.

On this page, you will find links to study guides for the exam, as well as information about registering for the exam. Although this book wasn't intended to be a study guide for the exam, it should get you most of the way there.

HashiCorp Cloud Platform

If you are ready to fire up services but you don't want to take on the operational management of the Consul system, I strongly recommend signing up for the **HashiCorp Cloud Platform**. All of that work setting up the server and securing it is done for you. Getting started is really easy, just by signing up at the following URL: `https://cloud.hashicorp.com/`.

Part of the process requires you to create your own organization, and once that is complete, you have access to the HashiCorp Cloud Platform.

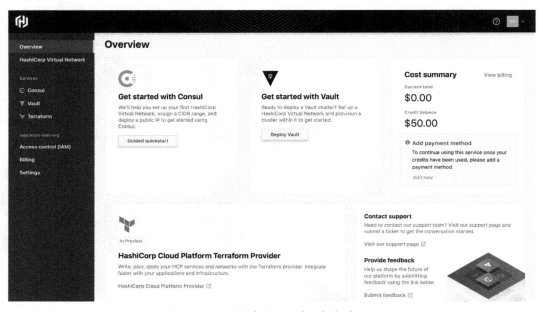

Figure 8.1 – HashiCorp Cloud Platform

The first thing we need to do is create a **HashiCorp Virtual Network**. This provides a network in which your HashiCorp cloud resources will operate.

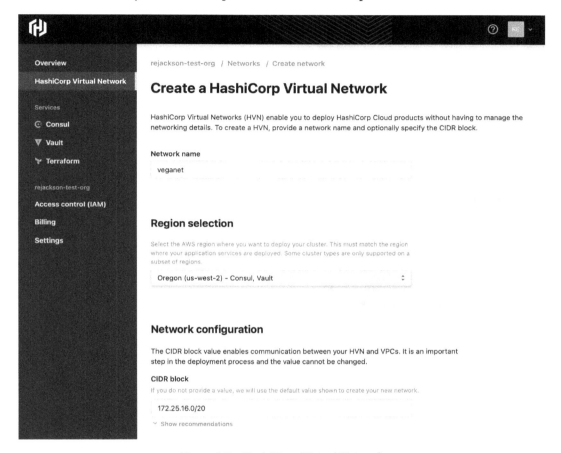

Figure 8.2 – HashiCorp Virtual Network

It takes a couple of minutes for the virtual network to be created, but once it is complete, we have the option to create a cluster. We have the option to deploy a Vault cluster, or a Consul cluster. As this is a book on Consul, and not Vault, let's go ahead and create a Consul cluster.

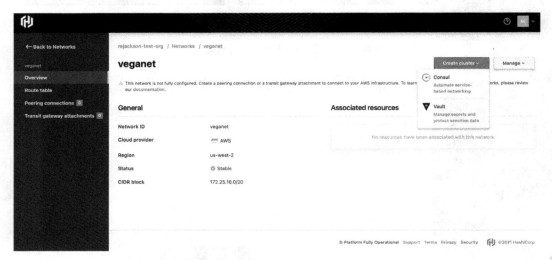

Figure 8.3 – HashiCorp Cloud Platform option to deploy a Consul cluster

Now we get to fill in the parameters needed to create the Consul cluster. There really isn't much to decide upon here, but starting with a small system and making it publicly accessible makes it a bit easier to connect. In production, special care should be taken regarding external connectivity to the cluster.

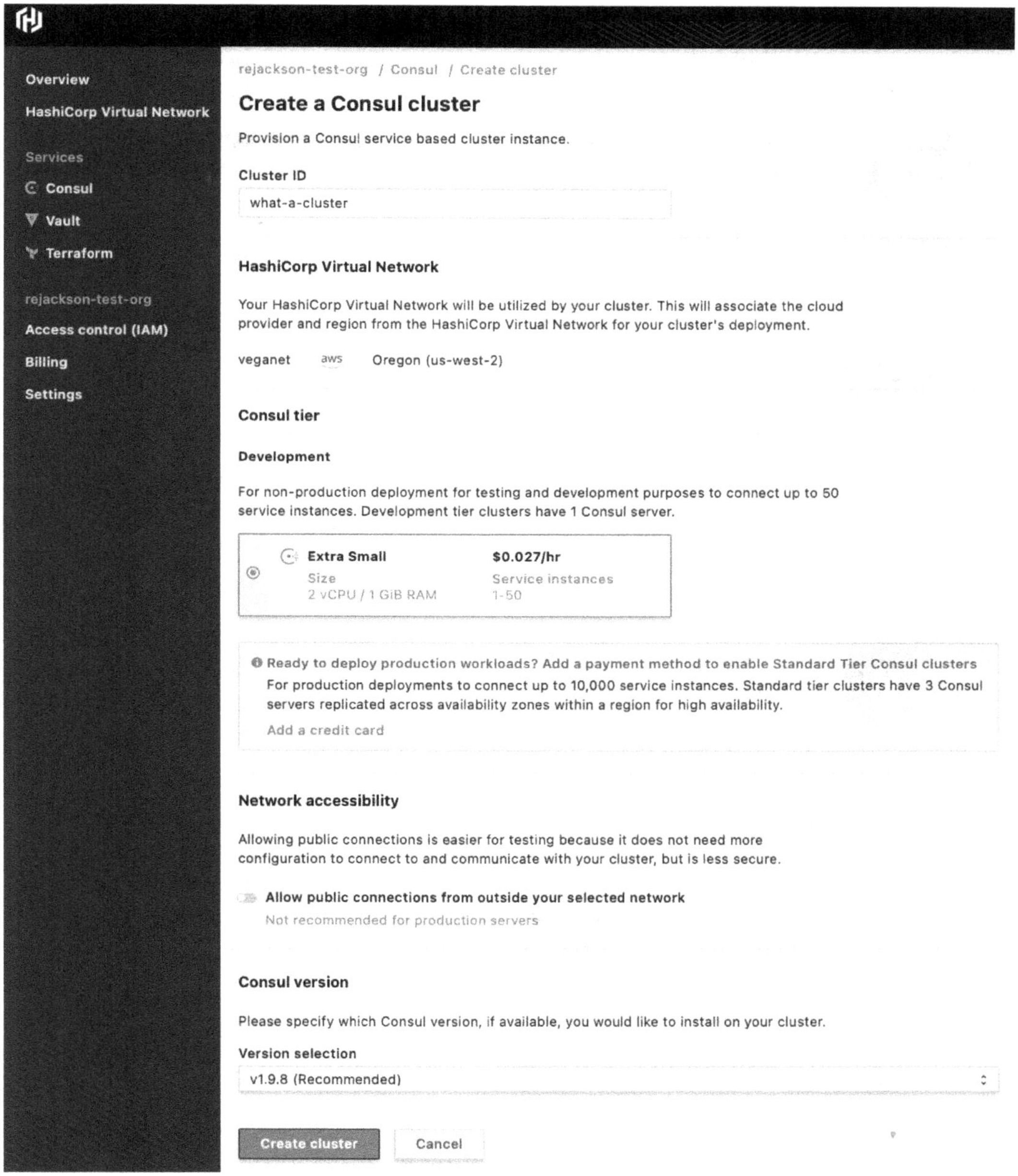

rejackson-test-org / Consul / Create cluster

Create a Consul cluster

Provision a Consul service based cluster instance.

Cluster ID

what-a-cluster

HashiCorp Virtual Network

Your HashiCorp Virtual Network will be utilized by your cluster. This will associate the cloud provider and region from the HashiCorp Virtual Network for your cluster's deployment.

veganet   aws   Oregon (us-west-2)

Consul tier

Development

For non-production deployment for testing and development purposes to connect up to 50 service instances. Development tier clusters have 1 Consul server.

	Extra Small	**$0.027/hr**
⦿	Size	Service instances
	2 vCPU / 1 GiB RAM	1-50

ⓘ Ready to deploy production workloads? Add a payment method to enable Standard Tier Consul clusters
For production deployments to connect up to 10,000 service instances. Standard tier clusters have 3 Consul servers replicated across availability zones within a region for high availability.
Add a credit card

Network accessibility

Allowing public connections is easier for testing because it does not need more configuration to connect to and communicate with your cluster, but is less secure.

Allow public connections from outside your selected network
Not recommended for production servers

Consul version

Please specify which Consul version, if available, you would like to install on your cluster.

Version selection

v1.9.8 (Recommended)

[Create cluster]   [Cancel]

Figure 8.4 – Configure a Consul cluster

Once you click that magic blue button, the HashiCorp Cloud Platform will start creating your very own Consul cluster, hosted on the cloud! Once the cluster is created, you'll be presented with a screen similar to the following screenshot:

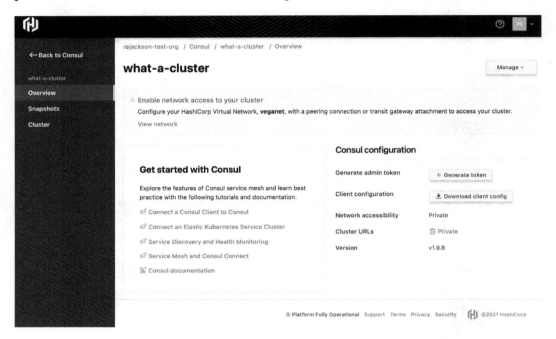

Figure 8.5 – Cluster Overview page in HashiCorp Cloud Platform

From here, you can generate your admin token, access the cluster URL in a web browser, and you're ready to go! In order to add clients, download the client configuration and distribute it to your Consul clients. The downloaded package includes the certificate that clients require to access the system, and any additional configuration necessary, and that's all you need to begin your journey into HashiCorp Consul in the cloud!

This is the end

My beautiful friend, hopefully not my only friend, this is the end. Or perhaps, and hopefully, it is the beginning of your journey with Consul. If you have a history with Consul, you may have noticed that there is a ton of information that we've missed, and I agree. Throughout the creation of this book, I continued to work my day job, identifying additional areas to cover. Kubernetes is an example that likely warrants an entire book on its own. However, I do hope that we've laid a foundation on which to apply multiple usages and applications of HashiCorp's Consul. The communication paths among Consul nodes, arguably one of the most important aspects of the architecture, has been explained in some detail. We've gone through the exercise of securing our Consul cluster and ensuring not only the encryption of communication, but the verification of identity. We don't want outsiders in our group and listening to our gossip, do we? Extending Consul across multiple data centers and regions is the next logical step for the architecture. Of course, once we have that stable foundation, we need to make sure we are getting value for our operational and capital investment. Consul provides that value through service discovery in our dynamic and ever-changing world. Simply finding our services in this world tends not to be so simple! Once we are able to locate our services, we utilize Consul to connect them across various environments, both internal and external, to Consul's domain. Of course, the networking world does extend far beyond Consul's domain, but we can apply Consul's knowledge to automate that network infrastructure. Lastly, although Consul can be deployed on nearly any system (some people run Consul on Raspberry Pis!), we often don't want to deal with operating systems if we don't have to. Consul on HashiCorp Cloud Platform provides the ability to quickly spin up a hosted Consul cluster and plug in your services.

I do hope you've enjoyed the book and learned a bit about Consul. More importantly, I hope that I've piqued your curiosity on a number of topics. Never stop learning!

Packt.com

Subscribe to our online digital library for full access to over 7,000 books and videos, as well as industry leading tools to help you plan your personal development and advance your career. For more information, please visit our website.

Why subscribe?

- Spend less time learning and more time coding with practical eBooks and Videos from over 4,000 industry professionals

- Improve your learning with Skill Plans built especially for you

- Get a free eBook or video every month

- Fully searchable for easy access to vital information

- Copy and paste, print, and bookmark content

Did you know that Packt offers eBook versions of every book published, with PDF and ePub files available? You can upgrade to the eBook version at packt.com and as a print book customer, you are entitled to a discount on the eBook copy. Get in touch with us at customercare@packtpub.com for more details.

At www.packt.com, you can also read a collection of free technical articles, sign up for a range of free newsletters, and receive exclusive discounts and offers on Packt books and eBooks.

Other Books You May Enjoy

If you enjoyed this book, you may be interested in these other books by Packt:

HashiCorp Infrastructure Automation Certification Guide

Ravi Mishra

ISBN: 978-1-80056-597-5

- Effectively maintain the life cycle of your infrastructure using Terraform 1.0
- Reuse Terraform code to provision any cloud infrastructure
- Write Terraform modules on multiple cloud providers
- Use Terraform workflows with the Azure DevOps pipeline
- Write Terraform configuration files for AWS, Azure, and Google Cloud
- Discover ways to securely store Terraform state files
- Understand Policy as Code using Terraform Sentinel
- Gain an overview of Terraform Cloud and Terraform Enterprise

Terraform Cookbook

Mikael Krief

ISBN: 978-1-80020-755-4

- Understand how to install Terraform for local development
- Get to grips with writing Terraform configuration for infrastructure provisioning
- Use Terraform for advanced infrastructure use cases
- Understand how to write and use Terraform modules
- Discover how to use Terraform for Azure infrastructure provisioning
- Become well-versed in testing Terraform configuration
- Execute Terraform configuration in CI/CD pipelines
- Explore how to use Terraform Cloud

Packt is searching for authors like you

If you're interested in becoming an author for Packt, please visit `authors.packtpub.com` and apply today. We have worked with thousands of developers and tech professionals, just like you, to help them share their insight with the global tech community. You can make a general application, apply for a specific hot topic that we are recruiting an author for, or submit your own idea.

Share Your Thoughts

Now you've finished *Simplifying Service Management with Consul*, we'd love to hear your thoughts! Scan the QR code below to go straight to the Amazon review page for this book and share your feedback or leave a review on the site that you purchased it from.

`https://packt.link/r/1800202628`

Your review is important to us and the tech community and will help us make sure we're delivering excellent quality content.

Index

A

Access Control Lists (ACLs)
  about 135
  working with 93-105
access control system, Consul
  controlling 68-76
ActiveMQ 152
admin-level privileges 75
anything as a service (*aaS) 176
API 148-150
Application layer 157
automation
  need for 182, 183
AWS credentials
  obtaining 20-26

B

bart.private_key certificate 66
bart.public_key certificate 66
binding rule 78
bootstrap policy 124

C

Certificate Authority (CA) 66, 160
checks implementing, service
  Docker 145
  gRPC 145
  H2ping 145
  HTTP 144
  script 143
  TCP 144
  TTL 145
claim mapping 77
cluster constituents 6, 7
communication paths
  calls, securing 65-68
  gossip, securing 63, 64
  protecting 63
connect functionality 174
consensus
  about 27
  among servers 28
  rafting 30
Consul
  access control system 68
  authentication, implementing 76-81

www.ingramcontent.com/pod-product-compliance
Lightning Source LLC
Chambersburg PA
CBHW060549060326
40690CB00017B/3650